Recent Studies in Sustainable Energy Harvesting

Recent Studies in Sustainable Energy Harvesting

Edited by **Ted Weyland**

LANRYE
INTERNATIONAL

New Jersey

Published by Clanrye International,
55 Van Reypen Street,
Jersey City, NJ 07306, USA
www.clanryeinternational.com

Recent Studies in Sustainable Energy Harvesting
Edited by Ted Weyland

International Standard Book Number: 978-1-63240-448-0 (Hardback)

This book contains information obtained from authentic and highly regarded sources. Copyright for all individual chapters remain with the respective authors as indicated. A wide variety of references are listed. Permission and sources are indicated; for detailed attributions, please refer to the permissions page. Reasonable efforts have been made to publish reliable data and information, but the authors, editors and publisher cannot assume any responsibility for the validity of all materials or the consequences of their use.

The publisher's policy is to use permanent paper from mills that operate a sustainable forestry policy. Furthermore, the publisher ensures that the text paper and cover boards used have met acceptable environmental accreditation standards.

Trademark Notice: Registered trademark of products or corporate names are used only for explanation and identification without intent to infringe.

Printed in the United States of America.

Contents

Preface

The recent studies and analyses in the field of sustainable energy harvesting are discussed in this insightful book. Early 21st century has witnessed many developments in the field of sustainable energy harvesting technologies. Since then, many such technologies have evolved, advanced and even been successfully developed into hardware models for preserving the operational lifetime of low power electronic devices like mobile gadgets, smart wireless sensor networks, etc. Energy harvesting is a technique that harvests renewable energy which is freely and easily available from the environment to recharge or to put used energy back into the energy storage devices without the inconvenience of disturbing, or even discontinuing the routine operation of the particular application. Due to the information, experience and understanding gained in the past few years, there have been some major developments in the research of sustainable EH technology. This book looks at various features of sustainable EH technology and its future implications.

The information contained in this book is the result of intensive hard work done by researchers in this field. All due efforts have been made to make this book serve as a complete guiding source for students and researchers. The topics in this book have been comprehensively explained to help readers understand the growing trends in the field.

I would like to thank the entire group of writers who made sincere efforts in this book and my family who supported me in my efforts of working on this book. I take this opportunity to thank all those who have been a guiding force throughout my life.

Editor

Part 1

Past and Present:
Mature Energy Harvesting Technologies

A Modelling Framework for Energy Harvesting Aware Wireless Sensor Networks

Michael R. Hansen, Mikkel Koefoed Jakobsen and Jan Madsen
Technical University of Denmark, DTU Informatics, Embedded Systems Engineering
Denmark

1. Introduction

A Wireless Sensor Network (WSN) is a distributed network, where a large number of computational components (also referred to as "sensor nodes" or simply "nodes") are deployed in a physical environment. Each component collects information about and offers services to its environment, e.g. environmental monitoring and control, healthcare monitoring and traffic control, to name a few. The collected information is processed either at the component, in the network or at a remote location (e.g. the base station), or in any combination of these. WSNs are typically required to run unattended for very long periods of time, often several years, only powered by standard batteries. This makes energy-awareness a particular important issue when designing WSNs.

In a WSN there are two major sources of energy usage:

- Operation of a node, which includes sampling, storing and possibly processing of sensor data.
- Routing data in the network, which includes sending data sampled by the node or receiving and resending data from other nodes in the network.

Traditionally, WSN nodes have been designed as ultra low-power devices, i.e., low-power design techniques have been applied in order to achieve nodes that use very little power when operated and even less when being inactive or idle. By adjusting the duty-cycle of nodes, it is possible to ensure long periods of idle time, effectively reducing the required energy.

At the network-level nodes are equipped with low-power, low-range radios in order to use little energy, resulting in multi-hop networks in which data has to be carefully routed. A classical technique has been to find the shortest path from any node in the network to the base station and hence, ensuring a minimum amount of energy to route data. The shortest path is illustrated in Fig. 1. Fig. 1(b) shows the circular network layout, where the base station is labelled N_x. Fig. 1(a) is a bar-chart showing the distance (y-axis) from a node to the base station, the x-axis is an unfolding of the circular network, placing the base station, with a distance of zero, at both ends.

The routing pattern of a node in this network is based upon the distance from a node (e.g. N_c) and its neighbours (N_b and N_d) to the base station. The node N_c will route to the neighbour with the shortest distance to the base station (in this case N_b). In practice, nodes close to the base station (e.g. N_a and N_g) will be activated much more frequently than those far away from

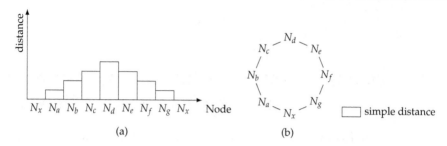

Fig. 1. An example network displaying the shortest distance to the base station. (a) shows each node's distance to the base station while (b) shows the placement of each node.

the base station, resulting in a relative short lifetime of the network. To address this, energy efficient algorithms, such as Bush et al. (2005); Faruque & Helmy (2003); Vergados et al. (2008), have been proposed. The aim of these approaches is to increase the lifetime of the network by distributing the data to *several* neighbours in order to minimize the energy consumption of nodes on the shortest path. However, these approaches do not consider the residual energy in the batteries. The energy-aware algorithms, such as Faruque & Helmy (2003); Hassanein & Luo (2006); Ma & Yang (2006); Mann et al. (2005); S.D. et al. (2005); Shah & Rabaey (2002); Xu et al. (2006); Zhang & Mouftah (2004), are all measuring the residual battery energy and are extending the routing algorithms to take into account the actual available energy, under the assumption that the battery energy is monotonically decreasing.

With the advances in energy harvesting technologies, energy harvesting is an attractive new source of energy to power the individual nodes of a WSN. Not only is it possible to extend the lifetime of the WSN, it may eventually be possible to run them without batteries. However, this will require that the WSN system is carefully designed to effectively use adaptive energy management, and hence, adds to the complexity of the problem. One of the key challenges is that the amount of energy being harvested over a period of time is highly unpredictable. Consider an energy harvester based on solar cells, the amount of energy being harvested, not only depends on the efficiency of the solar cell technology, but also on the time of day, local weather conditions (e.g., clouds), shadows from building, trees, etc.. For these conditions, the energy-aware algorithms presented above, cannot be used as they assume residual battery energy to be monotonically decreasing. A few energy harvesting aware algorithms have been proposed to address these issues, such as Islam et al. (2007); Lattanzi et al. (2007); Lin et al. (2007); Voigt et al. (2004; 2003); Zeng et al. (2006). They do not make the assumption of monotonically decreasing residual battery energy, and hence, can account for both discharging and charging the battery. Furthermore, they may estimate the future harvested energy in order to improve performance. However, these routing algorithms make certain assumptions that are not valid for multi-hop networks.

The clustering routing approach used in Islam et al. (2007); Voigt et al. (2004) assumes that all nodes are able to reach the base station directly. A partial energy harvesting ability is used in Voigt et al. (2003), where excess harvested energy can not be stored and the nodes are only battery powered during night. The algorithm in Lattanzi et al. (2007) is an offline algorithm, it assumes that the amount of harvestable energy can be predicted before deployment, which is not aa realistic assumption for most networks. The algorithm in Zeng et al. (2006) requires that each node have knowledge of its geographic position. Global knowledge is assumed in Lin et al. (2007).

Techniques for managing harvested energy in WSNs have been proposed, such as Corke et al. (2007); Jiang et al. (2005); Kansal et al. (2007; 2004); Moser et al. (2006); Simjee & Chou (2006). These are focussing on local energy management. In Kansal et al. (2007) they also propose a method to synchronise this power management between nodes in the network to reduce latency on routing messages to the base station. They do, however, not consider dynamic routes as such. An interesting energy harvesting aware multi-hop routing algorithm is the REAR algorithm by Hassanein & Luo (2006). It is based on finding two routes from a source to a sink (i.e. the base station), a primary and a backup route. The primary route reserve an amount of energy in each node along the path and the backup route is selected to be as disjunct from the primary route as possible. The backup route does not reserve energy along its path. If the primary route is broken (e.g. due to power loss at some node) the backup route is used until a new primary and backup route has been build from scratch by the algorithm. An attempt to define a mathematical framework for energy aware routing in multi-hop WSNs is proposed by Lin et al. (2007). The framework can handle renewable energy sources of nodes. The advantage of this framework is that WSNs can be analyzed analytically, however the algorithm relies on the ideal, but highly unrealistic assumption, that changes in nodal energy levels are broadcasted instantaneously to all other nodes. The problem with this approach is that it assumes global knowledge of the network.

The aim of this chapter is to propose a modeling framework which can be used to study energy harvesting aware routing in WSNs. The capabilities and efficiency of the modeling framework will be illustrated through the modeling and simulation of a distributed energy harvesting aware routing protocol, Distributed Energy Harvesting Aware Routing (DEHAR) by Jakobsen et al. (2010). In Section 2 a generic modeling framework which can be used to model and analyse a broad range of energy harvesting aware WSNs, is developed. In particular, a conceptual basis as well as an operational basis for such networks are developed. Section 3 shows the adequacy of the modeling framework by giving very natural descriptions and explanations of two energy harvesting based networks: DEHAR Jakobsen et al. (2010) and Directed Diffusion (DD) Intanagonwiwat et al. (2002). The main ideas behind routing in these networks are explained in terms of the simple network in Fig. 1. Properties of energy harvesting aware networks are analysed in Section 4 using simulation results for DEHAR and DD. These results validate that energy harvesting awareness increase the energy level in nodes, and hence, keep nodes (which otherwise would die) alive, in the sense that a complete drain of energy in critical nodes can be prevented, or at least postpone. Finally, Section 5 contains a brief summary and concluding remarks.

2. A generic modelling framework

The purpose of this section is to present a generic modelling framework which can be used to study energy-aware routing in a WSN, where the nodes of the network have an energy harvesting capability. In the next section instantiations of this generic model will be presented and experimental results through simulations are presented in Section 4.

The main idea of establishing a generic framework is to have a conceptual as well as a tool-based fundament for studying a broad range of wireless sensor networks with similar characteristics. In the following we will assume that

- sensor nodes have an energy-harvesting device,
- sensor nodes are using radio-based communication, consisting of a transmitter and a receiver,
- sensor nodes are inexpensive devices with limited computational power, and
- the routing in the network adapts to dynamic changes of the available energy in the individual nodes, i.e. the routing is energy aware.

On the other hand, we will not make any particular assumptions about the kind of sensors which are used to monitor the environment.

These assumptions have consequences concerning the concepts which should be reflected in the modelling framework, in particular, concerning the components of a node. Some consequences are:

- A node may only be able to have a direct communication with a small subset of the other nodes, called its neighbours, due to the range of the radio communication.
- A node needs information about neighbour nodes reflecting their current energy levels in order to support energy-aware routing.
- A node can make immediate changes to its own state; but it can only affect the state of other nodes by use of radio communication.
- The processing in the computational units as well as the sensing, receiving and transmitting of data are energy consuming processes.

These assumptions and consequences fit a broad range of WSNs.

The components of a node

A node consists of five physical components:

- An *energy harvester* which can collect energy from the environment. It could be by the use of a solar panel – but the concrete energy source and harvesting device are not important in the generic setting.
- A *sensor* which is used to monitor the environment. There may be several sensors in a physical node; but we will not be concerned about concrete kinds in the generic setting and will (for simplicity) assume that one generic sensor can capture the main characteristics of a broad range of physical sensors.
- A *receiver* which is used to get messages from the network.
- A *transmitter* which is used to send messages to the network.
- A *computational unit* which is used to treat sensor data, to implement the energy-aware routing algorithm, and to manage the receiving and sending of messages in the network.

The model should capture that use of the sensor, receiver, transmitter and computational unit consume energy and that the only supply of energy comes from the nodes' energy harvesters. It is therefore a delicate matter to design an energy-aware routing algorithm because a risk is that the energy required by executing the algorithm may exceed the gain by using it.

A consequence of this is that exact energy information cannot be maintained between nodes because it requires too much communication in the network as that would imply that too much energy is spent on this administrative issue compared to the harvested energy and the energy used for transmitting sensor-observations from the nodes to the base station.

The identity of a node

We shall assume that each node has a unique identification which is taken from a set Id of identifiers.

The state of a node

The *state* of a node is partitioned into a *computational state* and a *physical state*. The physical state contains a model of the real energy level in the node as well as a model of the dynamics of energy devices, like, for example, a capacitor. The computational state contains an approximation of the physical energy model, including at least an approximation of the energy level. The computational state also contains routing information and an abstract view of the energy level in neighbour nodes. Furthermore, the computational state could contain information needed in the processing of observations, but we will not go into details about that part of the computational state here, as we will focus on energy harvesting and energy-aware routing.

We shall assume the existence of the following sets (or types):

- PhysicalState – which models the real physical states of the node,

- Energy – which models energy levels,

- ComputationalState – which models the state in the computational unit in a node, including a model of the view of the environment (especially the neighbours) and information about the energy model and the processing of observations, and

- AbstractState – which models the abstract view of a computational state. An abstract state is intended to give a condensed version of a computational state and it can be communicated to neighbour nodes and used for energy-aware routing. It is introduced since it is too energy consuming to communicate complete state information to neighbours when radio communication is used.

The state parts of a node may change during operation. The concrete changes will not be described in the generic framework, where it is just assumed that they can be achieved using the functions specified in Fig. 2. Notice that a node can change its own state only.

Sets: PhysicalState, ComputationalState, AbstractState, and Energy

Operations:

consistent?	: ComputationalState $\rightarrow$ {true, false}
abstractView	: ComputationalState $\rightarrow$ AbstractState
updateEnergyState	: ComputationalState $\times$ Energy $\rightarrow$ ComputationalState
updateNeighbourView	: ComputationalState $\times$ Id $\times$ AbstractState $\rightarrow$ ComputationalState
updateRoutingState	: ComputationalState $\rightarrow$ ComputationalState
transmitChange?	: ComputationalState $\times$ ComputationalState $\rightarrow$ {true, false}
next	: ComputationalState $\rightarrow$ Id

Fig. 2. An signature for operations on the computational state

The intuition behind each function is given below. A concrete definition (or implementation) of the functions must be given in an instantiation of the generic model.

- consistent?(cs) is a predicate which is true if the computational state cs is *consistent*. Since neighbour and energy information, which are used to guide the routing, are changing dynamically, a node may end up in a situation where no neighbour seems feasible as the next destination on the route to the based station. Such a situation is called inconsistent, and the predicate consistent?(cs) can test for the occurrences of such situations.

- abstractView(cs) gives the abstract view of the computational state cs. This abstract view constitutes the part of the state which is communicated to neighbours.

- updateEnergyState(cs, e) gives the computational state obtained from cs by incorporation of the actual energy level e. The resulting computational state may be inconsistent.

- updateNeighbourView(cs, id, as) gives the computational state obtained from cs by updating the neighbour knowledge so that as becomes the abstract state of the neighbour node N_{id}. The resulting computational state may be inconsistent.

- updateRoutingState(cs) gives the computational state obtained from cs by updating the routing information on the basis of the energy and neighbour knowledge in cs so that the resulting state is consistent.

- transmitChange?(cs, cs') is a predicate which is true if the difference between the two computational states are so significant that the abstract view of the "new state" should be communicated to the neighbours.

- next(cs) gives, on the basis of the computational state cs, the identifier of the "best" neighbour to which observations should be transmitted.

The computation costs

Each of the above seven functions in Fig. 2 are executed on the computational unit of a node. Such an execution will consume energy and cause a change of the physical state. For simplicity, we will assume that the cost of executing the predicates consistent? and transmitChange? can be neglected or rather included in other functions, since they always incurs the same energy cost in these functions. These functions are specified in Fig. 3.

costAbstractView	: PhysicalState $\rightarrow$ PhysicalState
costUpdateEnergyState	: PhysicalState $\rightarrow$ PhysicalState
costUpdateNeighbourView	: PhysicalState $\rightarrow$ PhysicalState
costUpdateRoutingState	: PhysicalState $\rightarrow$ PhysicalState
costNext	: PhysicalState $\rightarrow$ PhysicalState

The costs of the predicates consistent? and transmitChange? are assumed negligible.

Fig. 3. An signature for cost operations on the computational state

For simplicity it is assumed that execution of each of the five functions have a constant energy consumption, so that all functions have the type PhysicalState $\rightarrow$ PhysicalState. It is easy to make this model more fine grained. For example, if the cost of executing abstractView depends on the computational state to which it is applied, then the corresponding cost function should have the type: PhysicalState $\times$ ComputationalState $\rightarrow$ PhysicalState. This level of detail is, however, not necessary to demonstrate the main principles of the framework.

Input events of a node

The computational unit in a node can react to *events* originating from the energy observations on the physical state, e.g. due to the harvesting device, the sensor and the receiver. There are two energy related events, where one is concerned with the change of the physical state while the other is concerned with reading the energy level in the node. The rationale for having two events rather than a "combined" one is that the change of the physical state is a cheap operation which does not involve a reading nor any other kind of computation, whereas a reading of the energy level consumes some energy.

A sensor recording results in an observation o belonging to a set Observation of observations. An observation could be temperature measurement, a traffic observation or an observation of a bird – but the concrete kind is of no importance in this generic part of the framework.

The events are described as follows:

- readEnergyEvent(e, ps), where $e \in$ Energy and $ps \in$ PhysicalState, which is an event signalling a reading e of the energy level in the node and a resulting physical state ps, which incorporates that the reading actually consumes some energy.

- physicalStateEvent(ps), where $ps \in$ PhysicalState is a new physical state. This event occurs when a change in the physical state is recorded. This change may, for example, be due to energy harvesting, due to a drop in energy level, or due to some other change which could be the elapse of time.

- observationEvent(o, ps), where $o \in$ Observation is a recorded sensor observation and $ps \in$ PhysicalState is a physical state which incorporates the energy consumption due to the activation of the sensor.

- receiveEvent(m, ps), where $ps \in$ PhysicalState and $m \in$ Message, which could be an observation to be transmitted to the base station or a message describing the state of a neighbour node. Further details are given below. The receiver maintains a *queue of messages*. When it records a new message, that message is put into the queue. The event receiveEvent(m, ps) is offered when m is the front element in the queue. Reacting to this event will remove m from the queue and a new receive event will be offered as long as there are messages in the queue. It is unspecified in the generic setting whether there is a bound on the size of the queue.

Input messages

A node has a queue of messages received from the network. There are two kinds of messages:

- *Observation Messages* of the form obsMsg(dst, o), where dst is the identity of the next destination of the observation $o \in$ Observation on the route to the base station.

- *Neighbour Messages* of the form neighbourMsg(src, as), where src is the identity of the source, i.e. the node which have sent this message, and $as \in$ AbstractState is the contents of the message in the form of an abstract state.

Let Message denote the set of all messages, i.e. observation and neighbour messages.

Output messages and communication

A node N_{id} can use the transmitter to broadcast a message $m \in$ Message to the network using the command $send_{id}(m)$. Intuitively, nodes which are within the range of the transmitter will receive this message and this may depend on the strength of the signal, it may depend on geographical positions, or on a variety of other parameters.

A model for sending and receiving messages could include a *global trace* of the messages send by nodes, a *local trace* of messages received by the individual nodes, and a description of a *medium*, that determines which nodes can receive messages sent by a node N_{id} on the basis of the current state of the network and on the basis of the various parameters, for example, concerning geographical positions of the nodes. In instances of the generic model, such a medium must be described. In this chapter we will not be formal about network communication. A formal model of communication along the lines sketched above can be found in Mørk et al. (1996); Pilegaard et al. (2003).

The cost of sending messages

Sending a message consumes energy which is reflected in a change of the physical state of a node. To capture this a function

$$\text{costSend} : \text{PhysicalState} \times \text{Message} \to \text{PhysicalState}$$

can compute a new physical state on the basis of the current one and a broadcasted message.

An operational model of a node

During its lifetime, a node can change between two main *phases*: *idle* and *treat message*.

- The node is basically inactive in the idle phase waiting for some event to happen. It processes an incoming event and makes a phase transition.

- The node treats a single message in the treat message phase and after that it makes a transition to the idle phase.

Each phase is parameterrised by the computational state cs and the physical state ps. The state changes and phase transitions for the idle phase are given in Fig. 4. The node stays inactive in the idle phase until a event occurs.

- A physical-state event leads to a change of physical state while staying in the idle phase.

- A read-energy event leads to an update of the energy and routing parts of the computational state, and the physical state is updated by incorporation of the corresponding costs. If the changes of the computational state are insignificant then these changes are ignored (so that the nodes have a consistent knowledge of each other) and just the physical state is changed. Otherwise, the abstract view of the new computational state is computed and send to the neighbours, and both the computational and the physical states are changed.

- An observation event leads to a computation of the next node (destination) to which the observation should be transmitted on the route to the base station, and a corresponding observation message is sent. The physical state is changed with the cost of computing the destinations and the cost of sending a message while staying in the idle phase.

$$
\begin{aligned}
\text{Idle}_{id}&(cs, ps) = \\
&\texttt{wait} \\
&\quad \text{physicalStateEvent}(ps') \;\rightarrow\; \text{Idle}_{id}(cs, ps') \\[4pt]
&\quad \text{readEnergyEvent}(e, ps') \;\rightarrow \\
&\qquad \texttt{let } cs' = \text{updateRoutingState}(\text{updateEnergyState}(cs, e)) \\
&\qquad \texttt{let } ps'' = \text{costUpdateEnergyState}(\text{costUpdateRoutingState}(ps')) \\
&\qquad \texttt{if } \text{transmitChange}?(cs, cs') \\
&\qquad \texttt{then let } m = \text{neighbourMsg}(id, \text{abstractView}(cs')) \\
&\qquad\qquad \text{send}_{id}(m); \text{Idle}_{id}(cs', \text{costSend}(\text{costAbstractView}(ps''), m)) \\
&\qquad \texttt{else } \text{Idle}_{id}(cs, ps'') \\[4pt]
&\quad \text{observationEvent}(o, ps') \;\rightarrow \\
&\qquad \texttt{let } dst = \text{next}(cs) \\
&\qquad \texttt{let } m = \text{obsMsg}(dst, o) \\
&\qquad \text{send}_{id}(m); \text{Idle}_{id}(cs, \text{costSend}(\text{costNext}(ps'), m)) \\[4pt]
&\quad \text{receiveEvent}(m, ps') \;\rightarrow\; \text{TreatMsg}_{id}(m, cs, ps')
\end{aligned}
$$

Fig. 4. The Idle Phase

- A receive event indicates a pending message in the queue. That message is treated by a transition to the treat message phase.

Notice that all phase transitions from the idle phase preserve the consistency of the computational state. The only non-trivial transition to check is that from $\text{Idle}_{id}(cs, ps)$ to $\text{Idle}_{id}(cs', \text{costSend}(\text{costAbstractView}(ps''), m))$. The consistency of cs' follows since $cs' = \text{updateRoutingState}(\text{updateEnergyState}(cs, e))$ and updateRoutingState is expected to return a consistent computational state, at least under the assumption that cs is consistent.

The state changes and phase transitions for the treat message phase are given in Fig. 5. In this phase the node treats a single message. After the message is treated a transition to the idle phase is performed, where it can react to further events including the receiving of another message. A message is treated as follows:

- An observation message is treated by first checking whether this node is the destination for the message. If this is not the case, a direct transition to the idle phase is performed. Otherwise, the next destination is computed, the observation is forwarded to that destination and the physical state is updated taking the computation costs into account. The energy consumed by the test whether to discard or process a message is included in the energy consumption for receiving a message.

- A neighbour message must cause an update of the neighbour view part of the computational state giving a new state cs'. A new routing state cs'' must be computed. If the changes to the computational state is insignificant (in the sense transmitChange?(cs, cs'') is false and cs' is consistent), then a transition to the idle phase is performed with a computational state that is just updated with the new neighbour knowledge, and the physical which is updated by the computation cost. Otherwise, an abstract view of the computational state must be communicated to the neighbours, and the computational and the physical states are updated accordingly.

$$
\begin{aligned}
&\mathrm{TreatMsg}_{id}(m, cs, ps) = \\
&\quad \texttt{case } m \texttt{ of} \\
&\qquad \mathrm{obsMsg}(dst, o) \;\rightarrow \\
&\qquad\quad \texttt{if } id = dst \\
&\qquad\quad \texttt{then let } dst' = \mathrm{next}(cs) \\
&\qquad\qquad\quad \texttt{let } m' = \mathrm{obsMsg}(dst', o) \\
&\qquad\qquad\quad \mathrm{send}_{id}(m'); \mathrm{Idle}_{id}(cs, \mathrm{costSend}(\mathrm{costNext}(ps), m)) \\
&\qquad\quad \texttt{else } \mathrm{Idle}_{id}(cs, ps)) \\
\\
&\qquad \mathrm{neighbourMsg}(src, as) \;\rightarrow \\
&\qquad\quad \texttt{let } cs' = \mathrm{updateNeighbourView}(cs, src, as) \\
&\qquad\quad \texttt{let } cs'' = \mathrm{updateRoutingState}(cs') \\
&\qquad\quad \texttt{let } ps' = \mathrm{costUpdateNeighbourView}(\mathrm{costUpdateRoutingState}(ps)) \\
&\qquad\quad \texttt{if } \mathrm{transmitChange?}(cs, cs'') \vee \neg\mathrm{consistent?}(cs') \\
&\qquad\quad \texttt{then let } as' = \mathrm{abstractView}(cs') \\
&\qquad\qquad\quad \texttt{let } m = \mathrm{neighbourMsg}(id, as') \\
&\qquad\qquad\quad \mathrm{send}_{id}(m); \mathrm{Idle}_{id}(cs'', \mathrm{costSend}(\mathrm{costAbstractView}(ps'), m)) \\
&\qquad\quad \texttt{else } \mathrm{Idle}_{id}(cs', ps')
\end{aligned}
$$

Fig. 5. The Treat-Message Phase

Notice that all phase transitions from the treat-message phase preserve the consistency of the computational state. The consistency preservation due to observation messages is trivial. The transition from $\mathrm{TreatMsg}_{id}(m, cs, ps)$ to $\mathrm{Idle}_{id}(cs'', \mathrm{costSend}(\mathrm{costAbstractView}(ps'), m))$ preserves consistency since cs'' is constructed by application of updateRoutingState, and this function is expected to return a consistent computational state. The transition from $\mathrm{TreatMsg}_{id}(m, cs, ps)$ to $\mathrm{Idle}_{id}(cs', ps')$ also preserves consistency since that transition can only occur when the if-condition $\mathrm{transmitChange?}(cs, cs'') \vee \neg\mathrm{consistent?}(cs')$ is false.

Some of the main features of the operational descriptions in Fig. 4 and Fig. 5 are:

- A broad variety of instances of the operational descriptions can be achieved by providing different models for the sets and operations in Fig. 2 and Fig. 3. This emphasizes the generic nature of the model.

- The energy and neighbour parts of the model appear explicitly through the occurrence of the associated operations. Hence it is clear that the model reflects energy-aware routing using neighbour knowledge, and it is postponed to instantiations of the model to describe how it works.

- The energy cost model appears explicit in the form of the cost functions including the cost of events.

- A node will send a local view of its state to the neighbours only in the case when a significant change of the computational state has happened, which is determined by the transmitChange? predicate. The adequate definition of this predicate is a prerequisite for achieving a proper routing, as it is not difficult to imagine how it could load the network and drain the energy resources, if minimal changes to the states uncritically are broadcasted.

- The model is not biased towards a particular energy harvester and it is not biased towards and particular kind of sensor observation.

The generic model is based on the existence of a description of the medium through which the nodes communicates. This medium should at least determine which nodes can receive a message send by a given node in a given state. It may depend on the available energy, the geographical position, the distance from the sender, and a variety of other parameters. Furthermore, the medium may be unreliable so that messages may be lost.

The model describes the operational behavior (including the dynamics of the energy levels in the nodes) for the normal operation of a network. It would be natural to extend the model with an initialization phase where a node through repeated communications with the neighbours are building up the knowledge of the environment needed to start normal operations, i.e. making observations and routing them to the base station. We leave out this initialization part in order to focus on energy harvesting and energy-aware routing.

3. Instantiating the modelling framework

In this section it will be demonstrated that the energy-aware routing protocol DEHAR Jakobsen et al. (2010) can be considered as an instance of the generic modelling framework presented in the previous section. In order to do so, meaning must be given to the sets and operations collected in Fig. 2 and Fig. 3. This will provide a succinct presentation of the main ideas behind DEHAR. Furthermore, we will show that the DD protocol Intanagonwiwat et al. (2002) can be considered a special case of DEHAR. Concrete experiments, based on a simulation framework, depends on descriptions of the medium. This will be considered in Section 4.

3.1 A definition of the states

The abstract state comprises:

- A *simple distance* $d \in \mathcal{R}_{\geq 0}$ to the base station. This is described by a non-negative real number, where larger number means longer distance.
- An *energy-aware adjustment* $a \in \mathcal{R}_{\geq 0}$ of the distance for the route to the base station, where a larger distance means less energy is available.

Hence an abstract state is a pair $(d, a) \in \text{AbstractState}$, where

$$\text{AbstractState} = \mathcal{R}_{\geq 0} \times \mathcal{R}_{\geq 0}$$

For an abstract state (d, a), we call $\text{dist}(d, a) = d + a$ the *energy-adjusted distance*.

The computational state comprises:

- A *simple distance* $d \in \mathcal{R}_{\geq 0}$ to the base station, like the simple distance of an abstract state.
- An *energy level* $e \in \text{Energy}$.
- An *energy-faithful adjustment* $f \in \mathcal{R}_{\geq 0}$ capturing energy deficiencies along the route to the base station.
- A table nt containing entries for the *abstract state of neighbours*. This is modelled by the type: $\text{Id} \rightarrow \text{AbstractState}$.

Hence a computational state is a 4-tuple $(d, e, f, nt) \in$ ComputationalState, where

$$\text{ComputationalState} = \mathcal{R}_{\geq 0} \times \text{Energy} \times \mathcal{R}_{\geq 0} \times (\text{Id} \to \text{AbstractState})$$

We shall assume that there is a function energyToDist : Energy $\to \mathcal{R}_{\geq 0}$ that converts energy to a distance so that less energy means longer distance.

The value energyToDist(e) provides a local adjustment of the distance to the base station by just taking the energy level in the node into account. The intension with the energy-faithful adjustment is that the energy deficiencies along the route to the base station is taken into account, and the energy-faithful part is maintained by the use of the neighbour messages.

The *energy adjustment of a computational state* is the sum of the converted energy and the energy-faithful adjustment:

$$\text{adjust}(d, e, f, nt) = \text{energyToDist}(e) + f$$

and the energy-adjusted distance of a computational state is:

$$\text{dist}(d, e, f, nt) = d + \text{adjust}(d, e, f, nt) = d + \text{energyToDist}(e) + f$$

where we overload the dist function to be applied to both abstract and computational states. Furthermore, dist(id), $id \in$ Id, is the distance of the abstract state of the neighbour node N_{id}.

The function next : ComputationalState $\to$ Id should give the neighbour with the shortest energy-adjusted distance to the base station, i.e. the "best" neighbour to forward an observation. Hence, next(d, e, f, nt) is the identity id of the entry $(id, as) \in nt$ with the smallest energy-adjusted distance to the base station, i.e. the smallest dist(as). If several neighbours have the smallest distance an arbitrary one is chosen.

A computational state cs is consistent if next(cs) has a smaller energy-adjusted distance than cs, i.e. dist$(cs) >$ dist$(\text{next}(cs))$, hence

$$\text{consistent?}(cs) = \text{dist}(cs) > \text{dist}(\text{next}(cs))$$

A node with a consistent computational state has a neighbour to which it can forward an observation. But if the state is inconsistent, then all neighbours have longer energy-adjusted distances to the base station and it does not make sense to forward an observation to any of these neighbours.

We illustrate the intuition behind the adjusted distance using the example network example from Fig. 1. If the energy level in node N_e of this network is decreased, then the distance of N_e to the base station is increased accordingly (by the amount energyToDist(e)) as shown in Fig. 6. All nodes are still consistent; but in contrast to the situation in Fig. 1, the node N_d (in Fig. 6) has just one neighbour (N_c) with a shorter energy-adjusted distance to the base station.

Consider now the situation shown in Fig. 7 with energy adjustments for the nodes N_f and N_g. These adjustments make the node N_e inconsistent, since its neighbours N_d and N_f both have energy-adjusted distances which are longer than that of N_e. In the shown situation it would make no sense for N_e to forward observations to its "best" neighbour, which is N_f, since N_f would immediately return that observation to N_e since N_e is the "best" neighbour of N_f.

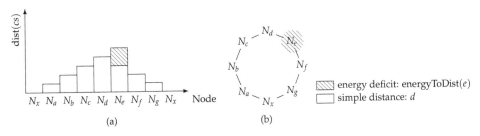

(a)   (b)

Fig. 6. The example from Fig. 1 with an energy adjustment for N_e due to shaded region shown to the right.

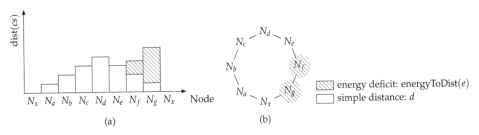

(a)   (b)

Fig. 7. Revised example with an inconsistent node: N_e.

Energy-faithful adjustments can be used to cope with inconsistent nodes. By adding such adjustments to the "problematic nodes" inconsistencies may be avoided. This is shown in Fig. 8, where energy-faithful adjustments (f) have been added to N_e and N_f. Every node is consistent, and there is a natural route from every node to the base station. From N_f there are actually two possible routes.

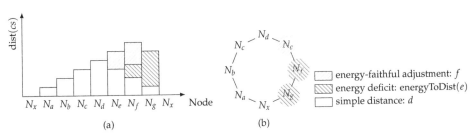

(a)   (b)

Fig. 8. A with consistent nodes using energy-faithful adjustments

The physical state comprises:

- The stored energy $e \in$ Energy.

- A model of the energy harvester. In the DEHAR case it is a solar panel, which is modelled by a function $P(t)$ describing the effect of the solar insolation at time t.

- A model of the energy store. In the DEHAR case it is an *ideal capacitor* with a given capacity. It is ideal in the sense that it does not loose energy.

- A model of the computational unit. This model must define the costs of the computational operations by providing definitions for the cost functions in Fig. 3. A simple way of doing this is to count the instructions needed for executing the individual functions, and multiply it with the energy needed per instruction. The model can be more fine grained by taking different modes of the processing unit into account.

- A model of the transmitter. This model must give a definition of the cost function: costSend : PhysicalState × Message → PhysicalState. In the DEHAR case the cost of sending is a simple linear function in the size of the message.

- A model of the receiver. This model must explain the cost of a receive event receiveEvent(m, ps). This involves the cost of receiving the message m and it must also take the intervals into account when the receiver is *idle listening*, i.e. it actively listens for incoming messages. Thus ps should reflect the full energy consumption of the receiver since the last receive event.

- A model of the sensor. This model must explain the cost of an observation event observationEvent(o, ps). This involves the cost of sensing o and ps should reflect this energy consumption.

The model should also describe two transitions of the physical state which relate to the two events physicalStateEvent(ps) and readEnergyEvent(e, ps).

The transition related to a physicalStateEvent must take into account at least the dynamics of the energy harvester, the dynamics of the energy store, the time the computational unit spent in the idle phase, and the time elapsed since the last physical state event. For example the new stored energy e' in the physical state at time t' is given by:

$$e' = e + \int_t^{t'} P(t)dt$$

where t is the time where the old energy e was stored.

The transition related to a readEnergyEvent(e, ps) must take into account at least the cost of reading the energy.

3.2 Definition of operations

The function for extracting the abstract view is defined by:

$$\text{abstractView}(d, e, f, nt) = (d, \text{adjust}(d, e, f, nt))$$

Notice that the distance to the base station is preserved by the conversion from a computational state to an abstract one:

$$\text{dist}(d, e, f, nt) = \text{dist}(\text{abstractView}(d, e, f, nt))$$

The definitions of the functions for updating the energy state and the neighbour view are simple:

$$\begin{aligned}
\text{updateEnergyState}((d, e, f, nt), e') &= (d, e', f, nt) \\
\text{updateNeighbourView}((d, e, f, nt), id, as) &= (d, e, f, \text{update}(nt, id, as))
\end{aligned}$$

where update(nt, id, as) gives the neighbour table obtained from nt by mapping id to the abstract state as. These two operations may transform a consistent state into an inconsistent one.

The function updateRoutingState(d, e, f, nt) must update the energy adjustment of a computational state in order to arrive at a consistent one. If the state is consistent even when $f = 0$ then no adjustment is necessary. Otherwise, an adjustment is made so that the distance of the computational state becomes K larger than the distance of its "best" neighbour (given by the next function):

$$\begin{aligned}
\text{updateRoutingState}(d, e, f, nt) = \\
\texttt{if } \text{consistent?}(d, e, 0, nt) \\
\texttt{then } (d, e, 0, nt) \\
\texttt{else let } \text{distNext} = \text{dist}(\text{next}(d, e, f, nt)) \\
(d, e, K + \text{distNext} - (d + \text{energyToDist}(e)), nt)
\end{aligned}$$

where $K > 0$ is a constant used to enforce a consistent computational state.

The energy adjustment in the `else`-branch of this function has the effect that the node becomes less attractive to forward messages to in the case of an energy drop in the node or in the best neighbour.

The function transmitChange?(cs, cs') is a predicate which is true when a change of the computational state from cs to cs' is significant enough to be communicated to the neighbours. This is the case if the change reflects a significant change in distance to base station, where significant in this case means larger than some constant $K_{change} \in \mathcal{R}_{\geq 0}$.

Hence, the function can be defined as follows:

$$\text{transmitChange?}(cs, cs') = |\text{dist}(cs) - \text{dist}(cs')| > K_{change}$$

A simple check of the operational descriptions in Fig. 4 and Fig. 5 shows that the new computational state used as argument to transmitChange? (cs' in Fig. 4 and cs'' in Fig. 5) must be consistent as it is created using updateRoutingState. Hence it is just necessary to define transmitChange? for consistent computational states.

Directed Diffusion – another instantiation of the generic framework

It should be noticed that the routing algorithm DD Intanagonwiwat et al. (2002) is a simple instance of the generic framework, which can be achieved by simplifying the DEHAR instance so that

- the simple distance is the number of hops to the based station (as for DEHAR) and
- the energy is assumed perfect and hence the adjustments have no effect (are 0).

Hence DD do not support any kind of energy-aware routing.

Actually, it is the algorithm behind DD which is used to initialize the simple distances of nodes in the DEHAR algorithm.

The DD algorithm provides a good model of reference for comparison with energy harvesting aware routing algorithms like DEHAR, since DD incorporates nodes with an energy harvesting capability, but the routing is static in the sense that an observation is always transmitted along the path with the smallest number of hops to the base station. Energy harvesting aware routing algorithms will not necessarily choose this shortest path, since problematic low-energy nodes should be avoided in order to keep all nodes "alive" as long as possible. Therefore, the total energy consumption in a DD based network should be smaller than the total energy consumption of any energy harvesting aware network (due to longer pathes in the latter). On the other hand, energy harvesting awareness can spare low-energy nodes, and there are two important consequences of this:

- A drain of low-energy nodes can be avoided or at least postponed. With regard to this aspect DD should perform worse since these nodes are not spared at all in the routing.
- The total energy stored in a network should exceed that of a corresponding DD based network, since messages are transmitted through nodes with good energy harvesting capability. The reason for this is that low-energy nodes get a chance to recover and that transmissions through high-energy nodes, with a full energy storage, are close to be "free of charge" since there would be almost no storage available for harvested energy in these high-energy nodes.

4. Results from simulation of the model

In this section we will study the properties of the energy harvesting aware routing algorithm DEHAR by analyzing results Jakobsen et al. (2010) of a simulator implementing the DEHAR and DD algorithms. The simulator is a custom-made simulator Jakobsen (2008) implemented in the language Java. It can be configured through a comprehensive xml configuration file which includes the network layout, environmental properties (insolation, shadows, etc.) and properties of nodes (such as processor states, radio model, and frequency of observations). The simulator features a classic event driven engine. The simulator produces a trace of observations of the nodes, including energy levels, activity of devices, and environmental properties

The considered network is given in Fig. 9. The network has one very problematic node, due to a strong shadow, at coordinate $(1,3)$, and five nodes with potential problems due to light shadows. We will analyse the ability of the routing algorithms to cope with these problematic nodes using simulations.

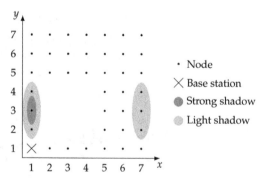

Fig. 9. A network structure with illustrating problematic nodes

The medium and the physical setting must be defined for the experiments. It is assumed that a node can communicate with its immediate horizontal and vertical neighbour, i.e. the radio range is 1. Two experiments S1 and S2 are conducted, one with a low and another with a high rate of conducted observations. Table 1 shows the parameters that are used in the presented simulations. Only the observation rate is changed between the two simulations.

			S_1	S_2	unit
Radio	Range		1	1	
	Transmit power		50	50	mW
	Idle listening power		5.5	5.5	mW
	Bandwidth		45	45	kb/s
Processor	Sleep	Power	1	1	µW
	Active	Frequency	1	1	MHz
		Power	10	10	µW
Battery	Capacity		4	4	kJ
Solar panel	Efficiency		6.25	6.25	%
	Area		12.5	12.5	cm^2
Application parameters	Observation rate		$\frac{1}{900}$	$\frac{1}{60}$	sec^{-1}
Routing parameters	Sense rate		$\frac{1}{1800}$	$\frac{1}{1800}$	sec^{-1}

Table 1. Parameters used in simulations.

The energy model is based on real insolation data for a two-weeks period. The data is repeated in simulations over longer periods. To emphasize the effect of the DEHAR algorithm, the insolation pattern have been idealised to either full noon or midnight, i.e. 12 hours of light and 12 hours of darkness. The insolation data is suitably scaled for individual nodes to achieve the shadow effect shown in Fig. 9.

Energy awareness makes a difference

A 30 day view of the simulations S_1 with the low observation rate is shown in Fig. 10. The figure shows the energy available in the worst node with minimum energy in the network. The two algorithms cannot be distinguished the first five days. Thereafter, the energy aware routing starts and DEHAR stabilises at a high level where no node is in any danger of being drained for energy. In the DD case, the energy of worst node is steadily drained at a (rather) constant rate and in an foreseeable future it will stop working.

Energy awareness consumes and stores more energy

The total power consumed and the average energy stored per node in the network are monitored for the same simulations as in Fig. 10. These results are shown for the first 10 days of simulated time in Fig. 11.

The day cycle is clearly visible in Fig. 11(a) where the nodes recharge during day and discharge during night. The first five days of simulation does not show any significant difference between DEHAR and DD. During the last five days the DEHAR algorithm makes the network able to harvest and store more energy.

The next graph (Fig. 11(b)) shows the difference of the two curves from the previous. It shows (in the blow-up) that just before day five ends, the DEHAR algorithm starts to consume

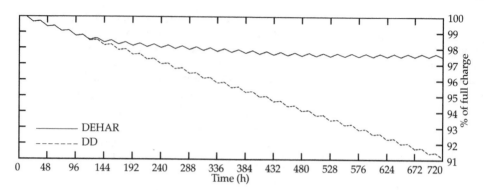

Fig. 10. Results of simulations S_1 for a 30 day simulation. This graph shows the minimum energy in any node in the network.

significantly more energy than the DD algorithm. By looking at the third graph (Fig. 11(c)) which shows the difference in total network energy consumption, it can be confirmed. This extra energy consumption arises from observation packages that travel along longer routes in the network, because the DEHAR algorithm have detected a lower amount of stored energy in some nodes.

Even though the DEHAR consumes more energy due to the longer routes, it can store more energy on average in the nodes. The reason for this is that the extra energy consumption of DEHAR is taken from nodes that are able to recharge fully during daytime. This can be seen in Fig. 11(b) (in the blow-up) at the beginning of day 5 (120h), where the graph shows a sudden rise.

After a short while, the network with the DD algorithm is able to harvest energy at a greater rate than DEHAR. This is due to the fact that the majority of the nodes in the DEHAR network are fully charged. The key point at this time is that the DD algorithm does not allow the network to harvest as much energy as the DEHAR algorithm. This can also be seen through the rest of the daylight during day 5, where the DEHAR network is able to harvest energy at a higher rate than the DD network.

Finally, during night, the DEHAR network again shows a higher energy consumption than the DD network. Hence the graph shows a slow decline.

Increasing the rate of observations costs

The next simulations (S_2) have an increased rate of observations and thus an increased radio traffic in the network. The effect of the increased data rate is primarily that the network consumes more power. This extra power consumption speeds up the time from the start of the simulation until the network finds the alternate routing pattern compared to the S_1 simulations.

Fig. 12 shows that the minimum energy in any of the nodes in the network stabilises with the DEHAR algorithm. The level at which it stabilises is lower than in the S_1 simulations, which is expectable. The faster observation rate hurts the DD network and a node will already be drained from energy in about 10 days.

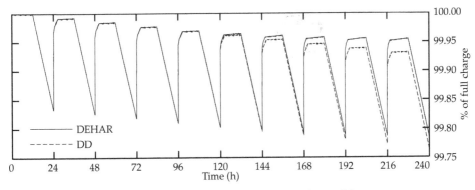

(a) Average energy in nodes for each simulation of S_1.

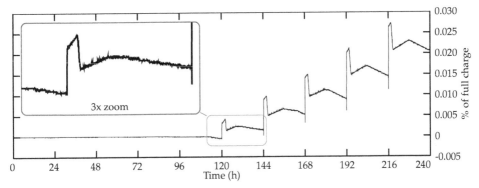

(b) Difference in the average energy in nodes for simulations in S_1. Given that the two curves in Fig. 11(a) are characterised by the functions $f_{DEHAR}(t)$ and $f_{DD}(t)$, then the curve in this figure is characterised by $f_{DEHAR}(t) - f_{DD}(t)$.

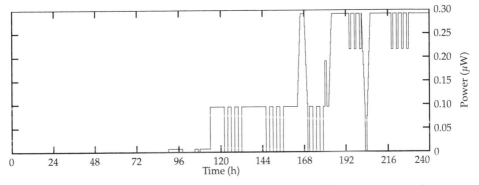

(c) Surplus energy consumption by DEHAR compared with DD for simulations in S_1.

Fig. 11. Results of simulations S_1 showing the first 10 days. The two blow-ups in (b) and (c) emphasises the first important difference between the DEHAR and DD algorithms.

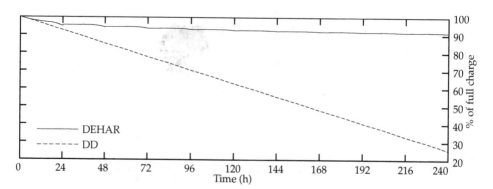

Fig. 12. Minimum energy in any node of the simulations in S_2. The day cycle is barely visible due to the compressed y-scale, compared to the simulations S_1.

The routing trend of the DEHAR algorithm is the same in the simulations S_1 and S_2. The only difference is that the DEHAR algorithm finds this alternative routing pattern faster in S_2 than in S_1.

The energy statistics of the node covered by the strongest shadow (at coordinate (1,3)) can be analysed. A graph of the energy level of this node will look similar to Fig. 12 and (in this simulation) it stabilises at precisely the same energy level. This show that the energy it can harvest closely matches the energy it needs to perform routing updates and performing observations (i.e. refraining from routing other nodes observations).

5. Conclusion

We have presented a new modelling framework aimed at describing and analysing wireless sensor networks with energy harvesting capabilities. The framework comprises of a conceptual basis and an operational basis, which were used to describe and explain two wireless sensor networks with energy harvesting capabilities. One of these network models is based on DD, i.e. it supports energy harvesting; but the routing is not energy aware, as it just forwards observations to the base station along statically defined shortest pathes. The other network model is based on the energy harvesting aware routing protocol DEHAR. Both of these networks were given natural explanations using the concepts of the modelling framework, and this gives a first weak validation of the adequacy of the framework. More experiments are, of course, needed for a thorough validation. Simulation results show that energy awareness of DEHAR-based networks can significantly extend the lifetime of nodes and it significantly improves the energy stored in the network, compared with a network like DD, with no energy aware routing.

There are several natural extension of this work.

First of all, the modelling framework should be validated by establishing its applicability in a broad collection of energy harvesting aware networks. The framework should be extended to include the deployment phase, where the nodes communicate in order to initialize their states. We do not expect principle difficulties in these extensions, but they are, of course, technical.

The generic framework may be instantiated in ways which will not be beneficial for the energy situation in the network. It is desirable and challenging to establish conditions which instantiations should satisfy in order to define an adequate energy harvesting aware network.

Another natural development would be to implement a platform for the modelling framework. The formalized parts of the framework provide good bases for such an implementation; but further formalization concerning the network communication and the medium should be considered prior to an implementation.

6. Acknowledgment

This research has partially been funded by the SYSMODEL project (ARTEMIS JU 100035) and by the IDEA4CPS project granted by the Danish Research Foundation for Basic Research.

7. References

Bush, L. A., Carothers, C. D. & Szymanski, B. K. (2005). Algorithm for Optimizing Energy Use and Path Resilience in Sensor Networks, *Wireless Sensor Networks, 2005. Proc. of the Second European Workshop on*, pp. 391 – 396.

Corke, P., Valencia, P., Sikka, P., Wark, T. & Overs, L. (2007). Long-duration solar-powered wireless sensor networks, *Proc. of the 4th workshop on Embedded networked sensors*, ACM, pp. 33–37.

Faruque, J. & Helmy, A. (2003). Gradient-based routing in sensor networks, *SIGMOBILE Mob. Comput. Commun. Rev.* 7(4): 50–52.

Hassanein, H. & Luo, J. (2006). Reliable Energy Aware Routing in Wireless Sensor Networks, *Dependability and Security in Sensor Networks and Systems, 2006*, IEEE, pp. 54–64.

Intanagonwiwat, C., Govindan, R., Estrin, D., Heidemann, J. & Silva, F. (2002). Directed Diffusion for Wireless Sensor Networking, *IEEE/ACM Transactions on Networking* 11(1): 2–16.

Islam, J., Islam, M. & Islam, N. (2007). A-sLEACH: An Advanced Solar Aware Leach Protocol for Energy Efficient Routing in Wireless Sensor Networks, *International Conference on Networking* 0: 4.

Jakobsen, M. K. (2008). *Energy harvesting aware routing and scheduling in wireless sensor networks*, Master's thesis, Technical University of Denmark, Department of Informatics and Mathematical Modeling.

Jakobsen, M. K., Madsen, J. & Hansen, M. R. (2010). DEHAR: A distributed energy harvesting aware routing algorithm for ad-hoc multi-hop wireless sensor networks, *World of Wireless Mobile and Multimedia Networks (WoWMoM), 2010 IEEE International Symposium on a*, pp. 1 –9.

Jiang, X., Polastre, J. & Culler, D. (2005). Perpetual environmentally powered sensor networks, *Information Processing in Sensor Networks, 2005. IPSN 2005. Fourth International Symposium on*, IEEE Press, pp. 463 – 468.

Kansal, A., Hsu, J., Zahedi, S. & Srivastava, M. B. (2007). Power management in energy harvesting sensor networks, *ACM Trans. Embed. Comput. Syst.* 6(4): 32.

Kansal, A., Potter, D. & Srivastava, M. (2004). Performance Aware Tasking for Environmentally Powered Sensor Networks, *ACM Joint Intl. Conf. on Measurement and Modeling of Computer Systems*.

Lattanzi, E., Regini, E., Acquaviva, A. & Bogliolo, A. (2007). Energetic sustainability of routing algorithms for energy-harvesting wireless sensor networks, *Comput. Commun.* 30(14-15): 2976–2986.

Lin, L., Shroff, N. B. & Srikant, R. (2007). Asymptotically optimal energy-aware routing for multihop wireless networks with renewable energy sources, *IEEE/ACM Transactions on Networking* 15(5): 1021–1034.

Ma, C. & Yang, Y. (2006). Battery-aware routing for streaming data transmissions in wireless sensor networks, *Mob. Netw. Appl.* 11(5): 757–767.

Mann, R. P., Namuduri, K. R. & Pendse, R. (2005). Energy-Aware Routing Protocol for Ad Hoc Wireless Sensor Networks, *EURASIP Journal on Wireless Communications and Networking* 2005(5): 635–644.

Mørk, S., Godskesen, J., Hansen, M. R. & Sharp, R. (1996). A timed semantics for sdl, *in* R. Gotzhein & J. Bredereke (eds), *Formal Description Techniques IX: Theory, application and tools*, Chapman & Hall, pp. 295–309.

Moser, C., Thiele, L., Benini, L. & Brunelli, D. (2006). Real-Time Scheduling with Regenerative Energy, *Proc. of the 18th Euromicro Conf. on Real-Time Systems*, IEEE Computer Society, pp. 261–270.

Pilegaard, H., Hansen, M. R. & Sharp, R. (2003). An approach to analyzing availability properties of security protocols, *Nordic Journal of Computing* 10: 337–373.

S.D., M., D.C.F., M., R.I., B. & A.O., F. (2005). A centralized energy-efficient routing protocol for wireless sensor networks, *IEEE Communications Magazine* 43(3): S8–13.

Shah, R. C. & Rabaey, J. M. (2002). Energy aware routing for low energy ad hoc sensor networks, *Wireless Communications and Networking Conf., 2002*, Vol. 1, IEEE, pp. 350 – 355.

Simjee, F. & Chou, P. H. (2006). Everlast: long-life, supercapacitor-operated wireless sensor node, *Proc. of the 2006 intl. symposium on Low power electronics and design*, ACM Press, pp. 197–202.

Vergados, D. J., Pantazis, N. A. & Vergados, D. D. (2008). Energy-efficient route selection strategies for wireless sensor networks, *Mob. Netw. Appl.* 13(3-4): 285–296.

Voigt, T., Dunkels, A., Alonso, J., Ritter, H. & Schiller, J. (2004). Solar-aware clustering in wireless sensor networks, *IEEE Symp. on Computers and Communications* 1: 238–243.

Voigt, T., Ritter, H. & Schiller, J. (2003). Solar-aware Routing in Wireless Sensor Networks, *Intl. Workshop on Personal Wireless Communications*, Springer, pp. 847–852.

Xu, J., Peric, B. & Vojcic, B. (2006). Performance of energy-aware and link-adaptive routing metrics for ultra wideband sensor networks, *Mob. Netw. Appl.* 11(4): 509–519.

Zeng, K., Ren, K., Lou, W. & Moran, P. J. (2006). Energy-aware geographic routing in lossy wireless sensor networks with environmental energy supply, *Proc. of the 3rd intl. conf. on Quality of service in heterogeneous wired/wireless networks*, ACM Press, p. 8.

Zhang, B. & Mouftah, H. T. (2004). Adaptive Energy-Aware Routing Protocols for Wireless Ad Hoc Networks, *Proc of the First Intl. Conf. on Quality of Service in Heterogeneous Wired/Wireless Networks*, IEEE Computer Society, pp. 252–259.

2

Modelling Theory and Applications of the Electromagnetic Vibrational Generator

Chitta Ranjan Saha

Score Project, School of Electrical & Electronic Engineering University of Nottingham,
Nottingham, NG7 2RD
UK

1. Introduction

There is rapidly growing interest over the last decade on the topics of energy harvesting devices as a means to provide an alternative to batteries as a power source for medical implants, embedded sensor applications such as buildings or in difficult to access or remote places where wired power supplies would be difficult [1-13]. There are several possible sources of ambient energy including vibrational, solar, thermal gradients, acoustic, RF, etc that can be used to power the sensor modules or portable electronic devices. The most promising ambient energy sources of these are solar, thermo-electric and vibrational. A significant amount of research has already been done in this area over the past few years and several energy scavenger products are already available in the market such as the solar calculator, thermoelectric wristwatch and wireless push button switches etc. The Solar energy is a mature technology and represents a very straight forward approach to generate energy from ambient light. However, solar cell is not cost effective and devices using solar cell need larger areas which would not be compatible with small MEMS powering. Furthermore sufficient sunlight is necessary which also limits the application areas. In thermoelectric generators, large thermal gradients are essential to generate practical levels of voltage and power. It would be very difficult to get more than 10°C in a MEMS compatible device. On the other hand, vibrational energy scavenger could be a reliable option for autonomous sensor modules or body-worn sensor, in automotive, industrial machine monitoring or other applications where ambient vibrational energy is available. This vibrational energy can be converted into electrical energy using three different principles: electromagnetic, electrostatic and piezoelectric.

The modelling theory of the electromagnetic (EM) vibrational generator (energy scavenger) and its applications are main objective in this chapter in order to understand the limitations of the EM energy harvesting device and how to increase voltage and power level for a specific application. Initially, this chapter gives the basic working principles of vibrational energy harvester and electrical machines. Then it will provide the modelling and optimization theory of the linear EM vibrational energy scavenger and discuss the analytical equations of each modelling parameter. Thereafter, this chapter presents the few macro scale cantilever prototypes which have been built and tested. Their measured results are discussed and analysed with the theory in order to see the accuracy of the model. It will also investigate the possible applications of the vibrational energy harvester. A prototype of the magnetic spring generator which has been built and tested for human body motion is presented and discussed the advantages of this structure. Finally we will present a prototype of optimized cantilever micro generator which has been built and integrated with the autonomous sensor module for machine monitoring application. The measured results of the real prototypes will provide the depth understanding of the readers what level of voltage and power could be harvested from the macro and micro level EM energy harvester and whether micro or macro device would be suitable for particular applications. The next section will give the brief overview of the working principle of the vibrational energy harvesters.

1.1 Kinetic/vibrational energy harvesting

Kinetic energy is the energy associated with the motion of an object. This includes vibrational motion, rotational motion and translational motion. The kinetic energy depends on two variables, the mass of the moving object (m) and the speed (U) of the object and is defined by [14];

$$K.E = \frac{1}{2}mU^2 \tag{1}$$

Kinetic energy is a scalar quantity and it is directly proportional to the square of its speed. In kinetic energy-harvesting, energy can be extracted from ambient mechanical vibrations using either the movement of a mass object or the deformation of the harvesting device. The basic operating principle of ac generator or alternator or EM harvester can be expressed using the energy flow diagram shown in Figure 1. When this external mechanical vibration or force is sufficient enough to overcome the mechanical damping force then the mass component of the energy harvesting devices to move or oscillate. This mechanical energy can be converted into electrical energy by means of an electric field (electrostatic), magnetic field (electromagnetic) or strain on a piezoelectric material, which are commonly known as electromechanical energy conversion principles. There also exists magnetostrictive energy harvesting devices which combine two principles: electromagnetic and piezoelectric.

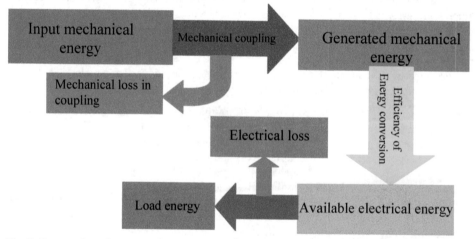

Fig. 1. Energy flow diagram of mechanical to electrical energy conversion principle.

Depending on the nature of the mechanical force, the generator can be classified in three categories: rotational generators, linear generators and deformation structure generators. The micro and macro scale linear or EM rotational generator, which is commonly known as an inertia generator in energy harvesting areas, will be investigated. Before introducing the EM energy harvester it is necessary to give a brief overview of the electrical machines such as transformer, motor and generator. Also the study of magnetic circuits is important since the operation of the EM energy harvester could be easily analyzed using the behavior of the magnetic fields. The next section will present the basic concepts of the electrical machines in order to understand the operating principle of the electromagnetic machines.

1.2 Concepts of electrical machine

An electrical machine is a electromechanical device that can convert either electrical energy to mechanical energy (known as a motor) or mechanical energy to electrical energy (known as generator). When such a device generates power in both directions it can be used as either a generator or a motor. The process of the electromechanical energy conversion normally involves the interaction of electric circuits and magnetic fields and the associated mechanical movement. This movement could be either rotational or linear due to forces arising between the fixed and the moving parts of the machine when we describe them as a rotational or linear machine. Another closely- related device is the transformer, which converts ac electrical energy at one voltage level to ac electrical energy at another voltage level. These three types of devices are very important in our everyday lives and sometimes such energy conversion devices are called transducer. One of the common factors between these machines is that they make use of magnetic fields to convert one form of energy to another. How these magnetic fields are used in such devices can be described by four basic principles [15-17];

1. A magnetic field will be produced surrounding a current-carrying conductor.
2. A time-changing magnetic field induces a voltage in a coil when it passes through it, which is called transformer action.
3. A current carrying conductor experiences a force in the presence of a magnetic field; this is known as motor action.
4. When a conductor such as copper wire moves in the magnetic field, a voltage will be induced between the conductor terminals; this is known as generator action.

The fourth principle is commonly known as Faraday's electromagnetic induction principle which has a wide range of applications, especially in power generation and power transmission theory. The following section will highlight the key components of the magnetic circuits since the magnetic field analysis is required to predict the performance of the electromagnetic device.

1.3 Magnetic materials and permanent magnet circuit model

Magnets are made from the magnetic materials and magnetic substances which consist of different metallic alloys. The magnetic materials are classified according to the nature of its relative permeability (μ_r) which is actually related to the internal atomic structure of the material and how much magnetization occurs within material. There are three categories the magnetic materials can be classified such as ferromagnetic materials, paramagnetic materials and diamagnetic materials. It is necessary to know the few quantities of the magnetic material such as magnetic flux density, B (T = wb/m²), the magnetizing force, H (A/m) and the magnetic flux, ϕ (wb). The relation between the magnetic flux density and the magnetizing force can be defined by;

$$B = \mu H = \mu_r \mu_0 H \tag{2}$$

Where μ (H/m) is the material permeability, μ_r is the relative permeability and μ_0 is the permeability in free space $4\,\pi$ x 10^{-7} H/m.

1.3.1 Ferromagnetic materials

The ferromagnetic materials have very large positive values of magnetic permeability and they exhibit a strong attraction to magnetic fields and are able to retain their magnetic properties after the external field has been removed. The relative permeability of ferromagnetic material could be a few hundred to a few thousand and they are highly nonlinear. Ferromagnetic materials those are easily magnetized called soft magnetic materials such as soft iron, silicon steel, soft ferrites, nickel-iron alloys etc. Soft magnetic materials have a steeply rising magnetization curve, relatively small and narrow hysteresis loop as shown in figure 2 (a). They are normally used in inductors, motors, actuators, transformer, sonar equipments and radars. Those ferromagnetic materials have a gradually rising magnetization curve, large hysteresis loop area and large energy loss for each cycle of magnetization as shown in figure 2 (b) called hard magnet or permanent magnet. Alnico, Ceramic, Rare-earth, Iron-chromium-Cobalt, Neodymium-Iron-boron etc are few examples of permanent magnet materials. The more details of the Hysteresis loop (B-H curve) is explained in different literatures [16-17].

1.3.2 Paramagnetic materials

The paramagnetic materials have small, positive values of magnetic permeability to magnetic fields. These materials are weakly attracted by the magnets when placed in a magnetic field and the materials could not retain the magnetic properties when the external field is removed. Potassium, aluminum, palladium, molybdenum, lithium, copper sulphate etc are common paramagnetic materials.

1.3.3 Diamagnetic materials

The diamagnetic materials have a weak, negative magnetic permeability to magnetic fields. Diamagnetic materials are slightly repelled by the magnets when placed in a magnetic field and the material does not retain the magnetic properties when the external field is removed. The examples of diamagnetic materials are bismuth, copper, diamond, gold etc.

Since the permanent magnet will be used to build the prototype of the electromagnetic vibrational power generator and it is necessary to understand the air gap flux density between magnet and coil. The magnetic excitation is supplied by permanent magnets which are used in all electromagnetic energy conversion devices and the air gap magnetic field density provides valuable information in evaluating the performance of any permanent magnet machine. Most designers use simplified analytical models of magnetic fields in the early stage design and then use FEA in the second stage of the design for better performance evaluation. For this work, rectangular- shaped magnets are assumed. A method for calculating the field at any point due to a rectangular magnet has been presented in [18]. Each magnet has eight corners and each corner flux density is equal to the remanent (B_r) of the magnet.

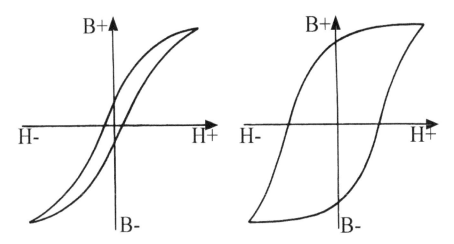

B : Magnetic flux density (T)

H : Ampere per meter/Magnetizing force

(a)   (b)

Fig. 2. Hysteresis loop (a) Soft magnetic materials (b) Hard magnetic materials.

Since the magnet has two opposite polarities, the remanent would be positive and negative consecutively as shown in figure 3. Suppose one corner co-ordinate is (x_k, y_k, z_k) with remanent $(-B_r)$ and the flux density at a point $M(x,y,z)$ outside the magnet can be calculated [18] from the following equation,

$$B_{zk} = -B_r \arctan(\frac{x_r y_r}{z_r \sqrt{x_r^2 + y_r^2 + z_r^2}}) \qquad (3)$$

Where B_{zk} is the flux density in the magnetization direction and $x_r = x - x_k$, $y_r = y_k - y$ and $z_r = z_k - z$. The total flux density $M(x,y,z)$ would be the summation of the eight corner flux densities.

Now we will discuss the magnetic circuit model of the permanent magnet. The second quadrant of the hysteresis loop is very valuable to describe the demagnetization characteristic of the permanent magnet. Figure 4 shows the linear demagnetization curve of the different permanent magnets [17-19]. Let consider the uniform cross sectional area and length of permanent magnet are A_p and l_p respectively as shown in figure 5. The intersection of the loop with the horizontal axis (H) is known as the coercive force, H_c and the vertical axis (B) is known as the remanent flux density (Br) or residual flux density. The linear demagnetization curve can be defined by;

$$B_p = \frac{B_r}{H_c}(H_p + H_c) = \mu_p(H_p + H_c) \tag{4}$$

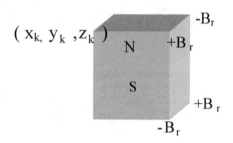

$$M(x,y,z)$$

Fig. 3. Magnet flux density calculation

Where $\mu_p = \dfrac{B_r}{H_c}$ is the permanent magnet permeability. The electrical analogy of the magnetic components for flux linkage (ϕ) as current (I), the magnetomotive force (F_p) as a voltage source (E) and the magnetic reluctance as the resistance. The voltage drop across the magnet can be defined as [19];

$$H_p l_p = \frac{l_p}{\mu_p A_p}\phi_p - H_c l_p = \Re_p \phi_p - F_p \tag{5}$$

Where $\Re_p$ is the magnetic reluctance of the permanent magnet, ϕ_p is the flux linkage and F_p is the magnetomotive force of the permanent magnet.

There is non linear demagnetization curve of the permanent magnet, linear magnetic circuit is still valid however the magnetic permeability would be;

$$\mu_p = \frac{B_p}{H_p + H_c} \tag{6}$$

The non linear magnetic circuit model of the permanent magnet can be represented in the figure 6. It can be seen from figure 7 that if the external magnetic field (H_{ex}) applied, the magnet operating point (H_p,B_p) would not move along the non-linear demagnetization curve, it simply moves along the centre line of the minor loop.

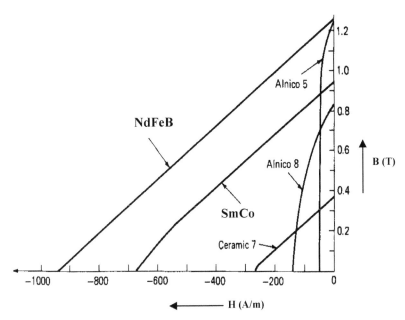

Fig. 4. Demagnetization curve of the permanent magnet [19]

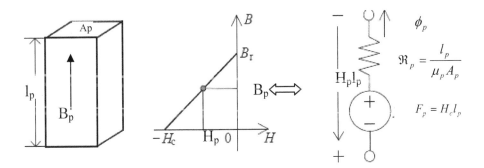

Fig. 5. Linear magnet circuit model of the permanent magnet.

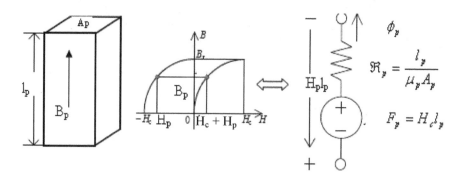

Fig. 6. Magnetic circuit model of a magnet using non-linear demagnetization curve.

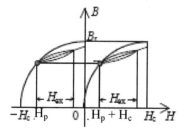

Fig. 7. Movement of the operating point of the magnet due to external magnetic field

In the next section, we will give a brief overview of Faraday's electromagnetic induction law and its fundamental equations.

1.4 Faraday's electromagnetic generator

Faraday discovered the electromagnetic induction which can be seen as a flux cutting phenomenon. If an electric conductor is moved through perpendicular to a magnetic field so as to cut the flux lines, a potential difference is induced between the ends of the conductor [16-17]. He also first developed the homopolar type electromagnetic generator which used a copper disc rotating between the poles of a horseshoe magnet; this is known as a Faraday disc. The principle of Faraday's law is applicable to the case where the time variation of voltage is created from the variation of the flux linkage of a stationary coil or a coil moves through a static magnetic flux or combination of both situations. Suppose the change in flux linkage in the circuit occurs within a small time interval Δt, then the induced emf can be defined by:

$$V = \frac{\Delta\phi}{\Delta t} = \frac{d\phi}{dt} \qquad (7)$$

where V is the generated voltage/induced emf and ϕ is the flux linkage.

If we consider the case where a coil moves in the x direction through a magnetic field or flux density B where B field varies along the coil movement, then the voltage can be expressed as:

$$V = \frac{d\phi}{dx}\frac{dx}{dt} \tag{8}$$

The flux linkage depends on the magnet and coil parameters and the air gap flux density between the magnet and coil. The shape of the air gap flux density could vary with the magnetic structure of the generator. The situation for vibrational generators can be approximated by a coil moving in a single direction through a magnetic field which varies in the direction of movement, as depicted in Figure 8. The flux linkage through a single turn conductor which encircles a surface area $(dA=dxdy)$, and which is positioned in a B field which varies with x but not y, can be expressed as;

$$\phi = \iint B.dA = \iint B.dxdy = \Delta y \int_{0}^{\Delta x} B(x)dx$$

and the flux linkage gradient is therefore;

$$\frac{d\phi}{dx} = \Delta y[B(\Delta x) - B(0)] \tag{9}$$

where $B(0)$ and $B(\Delta x)$ are the flux density at the $x=0$ position and the $x=\Delta x$ position.

The expression for the generated voltage as the product of the flux linkage gradient and the velocity is important for understanding the operation of the vibrational generator.

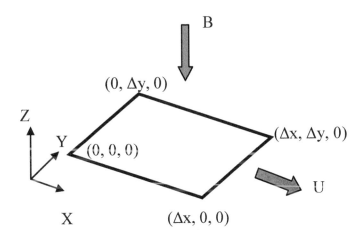

Fig. 8. Movement of a conductor in a position varying magnetic field.

1.4.1 Loudspeaker type vibrational generator

Figure 9 shows the schematic of a typical moving coil loudspeaker type linear generator. The loudspeaker type generator consists of a moving coil inside a static magnetic field where the voice coil is connected to the cone and its suspension [20]. Normally the cone is made from carbon fibre, plasticized cloth or paper. The magnet assembly consists of front plate, back plate, yoke pole pitch which are mainly made from low cost iron material and a large ferrite magnet. In the majority of loudspeaker designs the magnet height is longer than the coil height and the coil movement within the magnet height for full excursion. The flux density over the coil movement is constant for this kind of structure. The generated voltage would be;

$$V = BlU \tag{10}$$

Since the loudspeaker suspension is made from plastic, fibre or papers etc which are high loss material. Normally it is designed for small movement and will be flat response over the frequency range. The electromagnetic vibrational generator requires a particular frequency response with high efficiency.

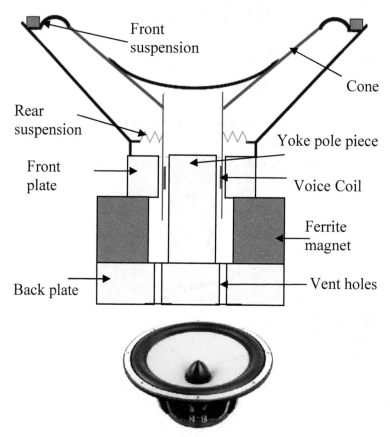

Fig. 9. Typical loudspeaker structure

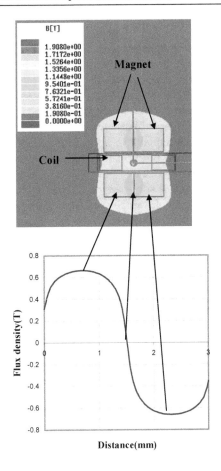

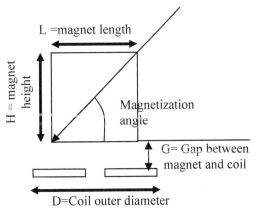

Fig. 10. Air gap flux density along the coil axis of the four magnets and single coil generator structure.

1.4.2 Four magnets vibrational generators

In this chapter rectangular shape four magnets and single coil generator structure is used for macro and micro scale prototype to verify the modeling theory. Figure 10 shows finite element simulation results of the air gap flux density (B) of the four magnet generator structure along the coil axis. This four magnet generator has several advantages. Since the coil is placed in the gap between upper and lower magnets, the field experienced by this coil is relatively large. In addition, the field is mostly perpendicular to the coil so that it contributes efficiently to the generated voltage. A four-magnets arrangement is used so as to produce two pole pairs. In its resting position, the coil is positioned mid-way between these pole pairs, i.e. half of the coil rests under one pole and half under the other. This is important as it means that the coil is subjected to a high flux gradient which should thus give rise to a relatively high voltage. It can be seen from this figure that the flux density is maximum at the mid position in each magnet and the flux density is zero between the mid points of the pole pairs.

It is fundamental to know the application before designing the generator. Need to consider the issues what level of acceleration or force and frequency are available and whether macro or micro size magnet and coil would generate sufficient power. Also what kind of generator structure such as magnet, coil and suspension would be suitable for the specific application? Finally how magnet and coil could be optimized to reduce cost and size for the specific application to deliver maximum power.

In the following section, we discuss the details of the model and the optimum condition for the generator.

1.5 Model of the electromagnetic vibrational generator

The basic vibrational energy harvester can be modeled as a second order mass damper and spring system [1-6]. It consists of a mass (m) mounted on a spring or a beam (k) which vibrates relative to a housing when subjected to an external vibrational force. For an electromagnetic generator mass may consist of the magnet itself or the coil. Figure 11 shows the typical structure of a generator as described by P.Glynne-Jones [21]. In this generator, two opposite polarity magnets were fixed in a gap at the free end of the cantilever and a wire-wound coil was placed in the gap between the magnets. When the external force is applied to the generator housing, the voltage would be generated in the coil terminal due to the relative displacement between magnets and coil. In order to have a model, it is important to develop the equations for the various generator parts such as the magnet parameters, the coil parameters, beam parameters and damping (D) parameters.

Figure 12 shows the schematic diagram of the electromagnetic generator. Variables z and y are the displacements of the generator mass and housing, respectively. It is assumed that movement of the housing is unaffected by the movement of the generator, since the moving mass (m) is much smaller than the mass of the generator housing. For a sinusoidal excitation, $y = Y_0 \sin(\omega t)$, where Y_0 is the vibration amplitude and ω is the frequency of vibration. The equation of motion for the mass relative to the housing at no load condition (no electromagnetic forces considered) can be defined [4-5] by the following equation,

Fig. 11. Typical generator structure

$$m\frac{d^2x}{dt^2} + D_p\frac{dx}{dt} + kx = -ma(t) = F_0\sin\omega t \tag{11}$$

where m is the moving mass of the generator, x is the relative movement between the mass and the housing, D_p is the parasitic damping, and $F_0 = m\omega^2 a$. The parasitic damping of the generator is commonly known as mechanical loss and can consist of air resistance loss, surface friction loss, material hysteresis loss, etc. It depends on material properties, the size and shape of the generator, external force, frequency, and vibrational displacement. k is the beam spring constant where the natural resonant frequency ω_n is given by, $\omega_n = \sqrt{k/m}$. The steady state solution of equation (4) is the displacement for the no-load condition and is given by the following equation [4]:

$$x_{no-load} = \frac{F_0\sin(\omega t - \phi)}{\sqrt{(k - m\omega^2)^2 + (D_p\omega)^2}} \text{ where, } \phi = \tan^{-1}(\frac{D_p\omega}{k - m\omega^2}) \tag{12}$$

This parasitic damping can be calculated from the open circuit quality factor and the damping ratio of the system, which can be expressed by;

$$Q_{oc} = \frac{m\omega_n}{D_p} , \quad \xi_{0c} = \frac{D_p}{2m\omega_n}$$

The damping ratio (ξ_{oc}) determines the qualitative behaviour of the system and it compares the time constant for decay of an oscillating system's amplitude to its oscillating period. The details of this parasitic damping and quality factor will be given later in this chapter.

It is well known that any mechanical, electrical or acoustic system always generates the maximum vibration amplitude at the resonance condition. Any system can have more than one resonance frequencies and resonance occurs when the system's natural frequency matches the frequency of oscillation of the external force.

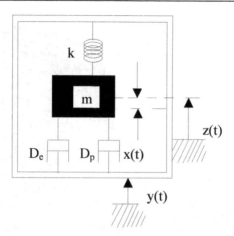

Fig. 12. Schematic representation of the vibrational generator.

The displacement at resonance ($\omega=\omega_n$) is given by;

$$x_{no-load} = \frac{-F_0 \cos(\omega_n t)}{D_p \omega_n} \tag{13}$$

and the phase angle, ϕ between displacement and the forcing signal is 90^0.

When a load is connected to the generator coil terminal, an electromagnetic force will be generated between the magnet and the coil due to the current flow through the load; this opposes the movement of the generator [5]. Thus, the equation of motion of the generator mass includes an extra term due to the magnetic force and becomes;

$$m\frac{d^2x}{dt^2} + D_p\frac{dx}{dt} + kx = F_0 \sin \omega t - F_{em} \tag{14}$$

where the F_{em} is the electromagnetic force.

The conductor moves along the X axis at velocity U in magnetic field B that varies with the position x as shown in Figure 2. In this case, the force experienced on the current-carrying conductors in the loop is;

$$F_{em} = \int IBdl = \int IB.dx = I[\int_{(0,0,0)}^{(\Delta x,0,0)} B.dx + \int_{(\Delta x,0,0)}^{(\Delta x,\Delta y,0)} Bdx + \int_{(\Delta x,\Delta y,0)}^{(0,\Delta y,0)} Bdx + \int_{(0,\Delta y,0)}^{(0,0,0)} Bdx]$$

$$F_{em} = I[\Delta x\{B(\Delta x) - B(0)\} + \Delta yB(\Delta x) + \Delta x\{B(0) - B(\Delta x)\} - \\ -\Delta yB(\Delta x)] = I\Delta y(B(0) - B(\Delta x))$$

Using equation (3).

$$F_{em} = I\frac{d\phi}{dx}$$

If the magnetic field B is constant with the position x then, $F_{em} = Bll$ where l is the coil mean length.

In this chapter we will present the magnetic flux density (B) varies with the coil movement, so that

$$F_{em} = \frac{V}{R_c + R_l + j\omega L}\frac{d\phi}{dx} = \frac{1}{R_c + R_l + j\omega L}(\frac{d\phi}{dx})(\frac{dx}{dt})(\frac{d\phi}{dx}) = \frac{(\frac{d\phi}{dx})^2}{R_c + R_l + j\omega L}\frac{dx}{dt}$$

where, V is the generated voltage, R_c is the coil resistance, L is the coil inductance, and R_l is the load resistance.

For an N turn coil, the total flux linkage gradient would be the summation of the individual turns flux linkage gradients. If the flux linkage gradient for each turn is equal then the electromagnetic force is given by;

$$F_{em} = \frac{N^2(\frac{d\phi}{dx})^2}{R_c + R_l + j\omega L}\frac{dx}{dt} = D_{em}\frac{dx}{dt}$$

Where the electromagnetic damping,

$$D_{em} = \frac{N^2(\frac{d\phi}{dx})^2}{R_c + j\omega L + R_l} \qquad (15)$$

It can be seen from (15) that electromagnetic damping can be varied by changing the load resistance R_c, the coil parameters (N, R_c and L), magnet dimension and hence flux (ϕ) and the generator structure which influence $\frac{d\phi}{dx}$. Putting the electromagnetic force

($F_{em} = D_{em}\frac{dx}{dt}$) in equation (7) gives;

$$m\frac{d^2x}{dt^2} + D_p\frac{dx}{dt} + D_{em}\frac{dx}{dt} + kx = F_0 \sin \omega t \qquad (16)$$

The solution of equation (9) defines the displacement under electrical load condition and is given by the following equation,

$$x_{load} = \frac{F_0 \sin(\omega t - \theta)}{\sqrt{(k - m\omega^2) + [(D_p + D_{em})\omega]^2}} \qquad (17)$$

Where $\theta = \tan^{-1}[\frac{(D_p + D_{em})\omega}{(k - m\omega^2)}]$

The displacement at resonance under load is therefore given by;

$$x_{load} = \frac{-F_0 \cos \omega t}{(D_p + D_{em})\omega} \tag{18}$$

1.5.1 Generated mechanical power

The instantaneous mechanical power associated with the moving mass under the electrical load condition is

$$P_{mech}(t) = F(t).U(t)$$

$$= F_0 \sin(\omega t) \frac{dx_{load}}{dt}$$

$$= \frac{F_0^2 \sin^2(\omega t)}{(D_p + D_{em})} \text{ using equation (18)}$$

Where $F(t)$ and $U(t)$ are the applied sinusoidal force and velocity of the moving mass, respectively, due to the sinusoidal movement. This corresponds to maximum mechanical power when $D_{em} = 0$, i.e. at no load.

The average mechanical power is defined by, $P_{mech} = \frac{1}{T} \int_0^T \frac{F_0^2 \sin^2(\omega t)}{(D_p + D_{em})} dt = \frac{F_0^2}{2(D_p + D_{em})}$

1.5.2 Generated electrical power and optimum damping condition

In a similar manner, the generated electrical power can be obtained from;

$$P_e(t) = F_{em}.U(t) = D_{em}U^2(t)$$

The average electrical power can be obtained from,

$$P_e = \frac{1}{T} \int D_{em} (\frac{dx_{load}}{dt})^2 dt \tag{19}$$

Taking the time derivative of equation (10) and putting the value in equation (12), we obtain

$$P_e = D_{em} \frac{(F_0 \omega)^2}{2[(k - m\omega^2)^2 + (D_p + D_{em})^2 \omega^2]} \tag{20}$$

The average electrical power generated at the resonance condition ($\omega = \omega_n$) is given by ;

$$P_e = D_{em} \frac{F_0^2}{2(D_p + D_{em})^2} \tag{21}$$

If the parasitic damping is assumed to be constant over the displacement range then the maximum electrical power generated can be obtained for the optimum electromagnetic damping. At the resonance condition $(\omega=\omega_n)$, the maximum electrical power and optimum electromagnetic damping can be found by setting $\dfrac{dP_e}{dD_{em}} = 0$ and solving for D_{em}. This gives the maximum power as;

$$P_{max} = \frac{F_0^{\,2}}{8D_p} \tag{22}$$

This occurs when $D_{em} = D_p$, which is the optimum electromagnetic damping at the resonance condition. Putting the value $D_{em} = D_p$ in equation (8) gives the displacement at the optimum load.

$$x_{load} = \frac{x_{no-load}}{2} \tag{23}$$

Thus, at the resonance condition, maximum power will be generated when the load displacement is half of the no-load displacement.

1.5.3 Maximum power and maximum efficiency

The maximum efficiency and maximum power depends on the external driving force and the design issues of the electromagnetic generators. If the driving force is fixed over the variation of the load and the electromagnetic damping or force factor (Bl) is significantly high compare to mechanical damping factor then the maximum power and maximum efficiency will appear at the same load resistance. Otherwise when the driving force is not constant and the force factor is significantly low or not high enough compare to mechanical damping or any of these situations the maximum power and the maximum efficiency will occur on different load resistance values.

1.5.4 Optimum load resistance for maximum generated electrical power

It is always desirable to operate the device at high efficiency and for an electrical generator, it is also desired to deliver maximum power to the load at a relatively high voltage. In an electromagnetic generator, most of the electrical power loss appears due to the coil's internal resistance. Here we will investigate what would be the optimum load resistance in order to get maximum power to the load. The electrical power and voltage lost in the coil internal resistance under these conditions are also investigated.

The optimum power condition occurs for $D_{em} = D_p$, which can be written as,

$$N^2 (\frac{d\phi}{dx})^2 \frac{1}{R_c + R_l + j\omega L} = D_p$$

In general, for less than 1 kHz frequency, $j\omega L$ can be neglected compared to R_c. Therefore, rearranging to get R_l, gives the optimum load resistance which ensures maximum generated electrical power namely,

$$R_l = \frac{N^2(\frac{d\phi}{dx})^2}{D_p} - R_c \qquad (24)$$

The above equation indicates that an optimum load resistance may not be positive if the first term on the right side is less than R_c . This can occur if either the parasitic damping factor (D_p) is large, the flux linkage gradient ($\frac{d\phi}{dx}$) is low, or the coil resistance is high. Since it is therefore not always possible to achieve the optimum condition by adjusting the load resistance, then it is worth considering the optimum conditions in various situations.

Very Low Electromagnetic damping case (Dem<<Dp) :

In the low electromagnetic damping case, due to low $\frac{d\phi}{dx}$ or high R_c , it is impossible to make the electromagnetic damping equal to the parasitic damping. If the electromagnetic damping for the short circuit condition is much less than the parasitic damping (D_{em}<<D_p), there will be no significant change in displacement between the no-load and load conditions. In this case, the maximum power will be delivered to the load when the load resistance is matched to the coil resistance. Since the load resistance has to be equal to the generator internal resistance, 50% of the voltage and power will be lost in the generator internal resistance and the generator efficiency is likely to be very low.

Limitation of the model ($D_p < D_{em}$<D_p) :

If the electromagnetic damping for very low load resistance is only slightly less than D_p, but can not be made equal to D_p then there will be a change in displacement between the no-load and load condition but the optimum load resistance at maximum generated power condition cannot be analyzed by the modeling equation. However, the optimum load resistance to maximize the load power condition, as opposed to the generated power could be determined from the modeling equation.

1.5.5 Optimum load resistance for maximum load power

In order to find the optimum resistance which maximizes the load power, we can take the expression for the load power and differentiate with respect to the load resistance.

The average generated electrical power is:

$$P_e = D_{em} \frac{F_0^2}{2(D_p + D_{em})^2}$$

The average load power would therefore be:

$$P_{load} = \frac{R_l}{R_c + R_l} [\frac{D_{em} F_0^2}{2(D_p + D_{em})^2}]$$

Inserting the expression for D_{em} from equation (15) and rearranging gives:

$$P_{load} = \frac{R_l F_o^2 N^2 (\frac{d\phi}{dx})^2}{2[D_p(R_c + R_l) + (\frac{d\phi}{dx})^2 N^2]^2}$$

Now the optimum load resistance at the maximum load power can be found by setting $\frac{dP_l}{dR_l} = 0$, which gives:

$$R_{lopt} = R_c + \frac{N^2 (\frac{d\phi}{dx})^2}{D_p} \tag{25}$$

In order to understand the optimum conditions of the generators, the displacement and load power were calculated theoretically for different parasitic damping factors. The parasitic damping, EM damping, displacement, generated voltage, the load power and the optimum load resistance at maximum load power were calculated by the following equations using the values in Table 1;

$$D_p = \frac{m\omega_n}{Q_{oc}} \; , \; D_{em} = N^2 \frac{(\frac{d\phi}{dx})^2}{R_c + R_l} \; , \; x_{load} = \frac{ma}{(D_p + D_{em})\omega_n} \; , \; V = \frac{d\phi}{dx}\frac{dx}{dt} \; ,$$

$$P_l = \frac{(VR_l)^2}{2R_l(R_c + R_l)^2} \; , \; R_{lopt} = R_c + \frac{N^2 (\frac{d\phi}{dx})^2}{D_p}$$

Parameters	value
N	500
Coil internal resistance, R_c (Ω)	33
Flux linkage gradient, $\frac{d\phi}{dx}$ (wb/m)	1e-03
Frequency, f (Hz)	1000
Acceleration, a (m/s²)	9.81
Mass (kg)	1.97e-03

Table 1. Assumed parameters of the Generators

Figure 13 shows the displacement vs load resistance, assuming different values of open circuit quality factor (Q_{oc}) for a 500 turns coil. It can be seen from the graphs that the significant variation of displacement for $Q_{oc} = 10000$ ($Dp=0.0012$ N.s/m) is due to the change in the load resistance value. Figures 14, 15 and 16 show the corresponding load power and damping factor vs load resistance. For $Q_{oc}=10000$, the maximum power is generated and transferred to the load when the electromagnetic damping is equal to the parasitic damping; this agrees with the theoretical model. In this case, the electromagnetic damping for very low load resistance is almost 6 times higher ($D_{em} \gg D_p$) than the parasitic damping factor. Since the optimum R_{load} is much greater than R_{coil}, 90% of the generated electrical power is delivered to the optimum load resistance value. The optimum load resistance at maximum load power is 255 Ω, which agrees with the theoretical equation (25).

For Q_{oc}=1000 there is some variation of displacement for changing load resistance values but it is not as significant as for the Q_{oc}=10000 case. In this case, electromagnetic damping for low load resistance is lower than parasitic damping ($D_{em} < D_p$) but not significantly lower. In this situation the optimum condition for the generated maximum power could not be defined by the modeling equation but the optimum load resistance at maximum load power is 55 which agrees well with theoretical equation (25). The optimum load resistance tends to be close in value to the coil resistance.

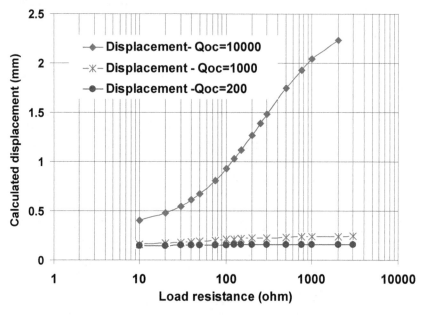

Fig. 13. Variation of displacement for different quality factors for N =500 turns coil generator

For Q_{oc}=200, there is no variation of displacement for changing load resistance values and the electromagnetic damping for all load resistances is significantly lower than the parasitic damping factor ($D_p \gg D_{em}$). In this case the maximum power is delivered to the load when the load resistance equals the coil resistance. It is assumed in the above that the parasitic damping is almost constant with the displacement. However, this parasitic damping can depend on the generator structure and the properties of the spring material such as friction, material loss etc.

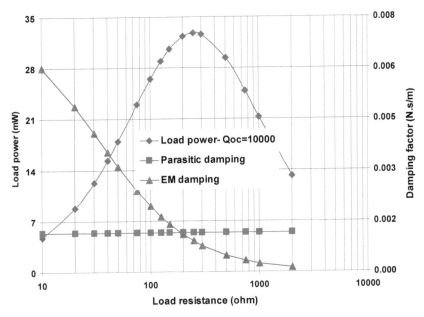

Fig. 14. Calculated load power and damping factor for $Q_{oc} = 10000$ and $N=500$ turns coil generator.

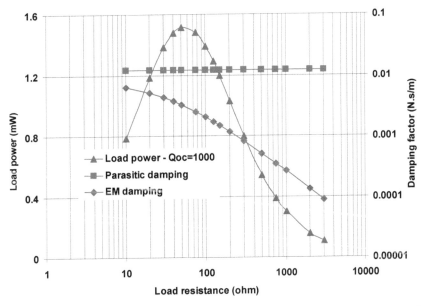

Fig. 15. Calculated load power and damping factor for $Q_{oc} = 1000$ and $N=500$ turns coil generator.

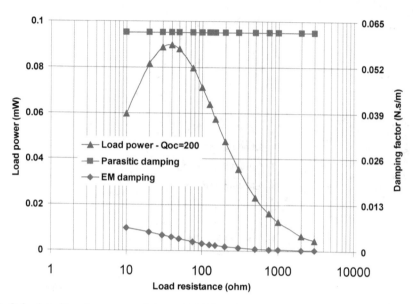

Fig. 16. Calculated load power and damping factor for Q_{oc} = 200 and N=500 turns coil generator.

1.6 Parasitic or mechanical damping and open circuit quality factor

The parasitic damper model of the mechanical beam in the electromagnetic vibrational generator structure is considered as a linear viscous damper [22-27]. The parasitic damping therefore determines the open-circuit or un-loaded quality factor which can be expressed as;

$$Q_p = \frac{m\omega_n}{D_p} = \frac{1}{2\varsigma_p} = \frac{f_n}{f_2 - f_1} \tag{26}$$

Where ς_p is the parasitic damping ratio, f_1 is is the lower cut-off frequency, f_2 is the upper cut-off frequency and f_n is the resonance frequency of the power bandwidth curve which is shown in graph 17. The quality factor can also be calculated from the voltage decay curve or displacement decay curve for the system when subjected to an impulse excitation, according to equation [22]:

$$Q_p = \frac{\pi f_n \Delta t}{\ln(\frac{x_1}{x_2})} = \frac{\pi f_n \Delta t}{\ln(\frac{V_1}{V_2})}$$

In general, the unloaded quality factor of a miniature resonant generator is influenced by various factors. At its most general, it can be expressed as:

$$Q_p = \left(\frac{1}{Q_m} + \frac{1}{Q_t} + \frac{1}{Q_c} + \frac{1}{Q_{su}} + \frac{1}{Q_f} \right)^{-1} \tag{27}$$

where $1/Q_m$ is the dissipation arising from the material loss, $1/Q_t$ is the dissipation arising from the thermoelastic loss, $1/Q_c$ is the dissipation arising from the clamping loss, $1/Q_{su}$ is the dissipation arising from the surface loss, and $1/Q_f$ is the dissipation arising from the surrounding fluid. There have been considerable efforts to find analytical expressions for these various damping mechanisms, particularly for Silicon-based MEMS devices such as resonators. However further analysis of the parasitic damping factor is beyond in this chapter.

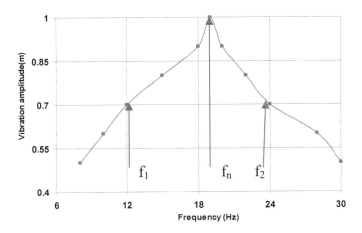

Fig. 17. Power bandwidth curve

1.7 Spring constant (k) of a cantilever beam

A cantilever is commonly defined as a straight beam, as shown in Figure 18 with a fixed support at one end only and loaded by one or more point loads or distributed loads acting perpendicular to the beam axis. The cantilever beam is widely used in structural elements and the equations that govern the behavior of the cantilever beam with a rectangular cross section are simpler than other beams. This section shows the equations that the maximum allowable vertical deflection, the natural frequency and spring constant due to the end loading of the cantilever.

The maximum allowable deflection the spring can tolerate is [27]:

$$y_{max} = \frac{2L^2}{3Et}\sigma_{max} \tag{28}$$

where σ_{max} is the maximum stress, E is Young's modulus, t is the thickness of the cantilever, and L is the length of the cantilever.

The maximum stress can be defined as

$\sigma_{max} = \dfrac{FLt}{2I}$ where F is the vertical applied force and $I = \dfrac{Wt^3}{12}$ is the moment of inertia of the beam.

The ratio between the force and the deflection is called the spring constant, k and is given by:

$$k = \frac{3EI}{L^3}$$

(29)

The total end mass of the beam is $m = 0.23M + m_1$

where m_1 is the added mass and M is the mass of cantilever.

The equation of motion for free undamped vibration is:

$m\frac{\partial^2 y}{\partial t^2} + ky = 0$, where, m is the total end mass of the beam. If $y = A\cos\omega_n t$ then the natural frequency would be,

$$f_n = \frac{1}{2\pi}\sqrt{\frac{k}{m}}$$

(30)

The next section presents the electrical circuit analogy of the electromagnetic vibrational generator.

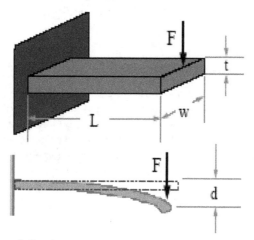

Fig. 18. Cantilever beam deflection

1.8 Equivalent electrical circuit of electromagnetic vibrational generator

The vibrational generator consists of mechanical and electrical components. The mechanical components can be easily represented by the equivalent electrical circuit model using any electrical spice simulation software in order to understand their interactions and behaviours. Two possible analogies either impedance analogy or mobility analogy is normally used in the transducer industry which compare mechanical to electrical systems. However it is good idea to use the analogy that allows for the most understanding and also it is easy to switch one analogy to other. Table 2 [29-30] shows the equivalent electrical circuit elements of the mechanical components for the electromagnetic vibrational generator. Figure 19 shows the equivalent electrical circuit of the force driven electromagnetic linear or vibrational generator. When an external force (F) is applied to the generator housing or diaphragm the voltage will be generated at the coil terminal due to the relative displacement between magnet and coil. The moving mass, mechanical compliance, mechanical resistance, coil resistance, coil inductance, load resistance and the force factor of the alternator are M_m, $C_m = 1/k$, R_m, R_c, L_c, R_L and Bl respectively.

Mechanical elements	Equivalent Electrical elements	
	Impedance analogy	Mobility analogy
Force (F)	Voltage (E)	Current (I)
Velocity (U)	Current (I)	Voltage (E)
Mechanical resistance (R_m)	Resistance (R)	Conductance (G)
Mechanical mass (M)	Inductance (L)	Capacitance (C)
Mechanical compliance (C_m)	Capacitance (C)	Inductance (L)

Table 2. Electrical equivalent of the mechanical components.

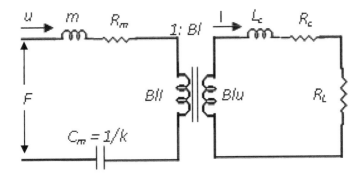

Fig. 19. Equivalent electrical circuit of the electromagnetic vibrational generator.

2. Verification of the model

In order to verify the model and the optimization theory several macro generators had been built and tested using a controllable shaker during Author's Ph.D study. Some of these works have already been highlighted in literatures [6]. It was important to vibrate the generator exactly at resonance in order to observe the electromagnetic damping effect for different load conditions. If the vibration frequency is far away from the system's resonant frequency, then the damping would not have any significant effect on displacement since the displacement is mainly controlled by the spring constant at off resonance. All the prototypes which have been built and tested consisted of four magnets (NdFeB35) with a wire-wound copper coil placed between the magnets as shown in Figure 20. The advantages of the four magnet generator structure have already been described in the previous section. Table 3 gives the coil and magnet parameters for macro generator A and B. The generators were vibrated using a sinusoidal acceleration with the frequency matched to the generator's mechanical resonant frequency.

Fig. 20. Generator A-showing four magnets attached to a copper beam and wire-wound coil.

Parameters	Generator -A	Generator-B
Moving mass (kg)	0.0428	0.025
Magnet size (mm)	15 x 15 x 5	10 x 10 x 3
Resonant frequency (Hz)	13.11	84
Acceleration (m/s²)	0.78	7.8
Magnet and coil gap (mm)	1.25	1.5
Coil outer diameter (mm)	28.5	13.3
Coil inner diameter (mm)	5	2
Coil thickness (mm)	7.5	7
Coil turns	850	300
Coil resistance (ohm)	18	3.65

Table 3. Generator Parameters.

Measured and calculated results of the macro-generator A & B

The displacement and voltage were measured for various load resistances. The load power is calculated from the voltage and load resistance. The parasitic damping can be calculated using the no-load displacement equation:

$$x_{no-load} = \frac{-F \cos \omega t}{D_p \omega} \; , \; D_p = \frac{F_0}{x_{no-load} \, \omega}$$

The generators were also simulated using a 3-D finite element (FE) transient model with the measured displacement as input. Figure 21 shows the half model of the generator structure used in the FE model and the simulated flux linkage vs displacement of macro generator A

& B. The simulated flux linkage gradients $(\frac{d\phi}{dx})$ of the generator A and B are 0.00542 Wb/m and 0.0013 Wb/m respectively.

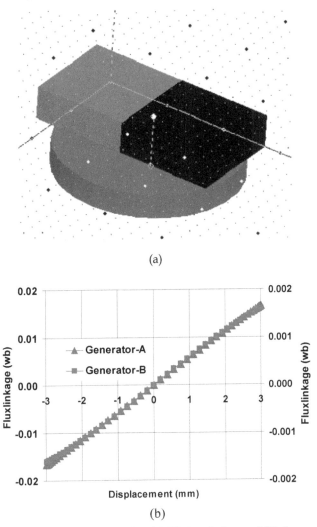

(a)

(b)

Fig. 21. Simulated Generator models used in (a) FE simulation and (b) the resulting flux linkage gradient vs displacement.

These values are used in the following equation to calculate the electromagnetic damping;

$$D_{em} = \frac{N^2 (\frac{d\phi}{dx})^2}{R_c + R_l}$$

 The parasitic and electromagnetic damping can then be used in the following equation to calculate the power:

$$P_{avg} = D_{em} \frac{F^2}{(D_p + D_{em})^2}$$

Figures 22 and 23 show the measured displacement and the measured and simulated load voltages for different load conditions for generator-A and generator B, respectively. The measured and simulated voltages agree quite closely. Figure 24 and 25 show the measured and calculated power, and the estimated parasitic and electromagnetic damping, for generators A and B, respectively. The calculated open circuit quality factors $Q_{oc} = \frac{m\omega_n}{D_p}$ are

58.85 and 56.87 for generators A & B, respectively. The graph in Figure 22 shows that there is a significant change in displacement with the change in load, for generator A. However for generator B, the displacement does not vary with load. This is consistent with the fact that the electromagnetic damping is comparable to the parasitic damping for generator A, but the electromagnetic damping is much lower for generator B than the parasitic damping due to larger gap between magnet and coil, and smaller magnets used. This can be seen from figure 24 and 25. Furthermore, the graph in figure 24 shows that the power readies a maximum for the value of load resistance at which electromagnetic damping and parasitic damping are equal. The optimum load resistance is 432 Ω, which agrees well with the theoretical equation. For generator B, the electromagnetic damping is always much less than the parasitic damping and the power is maximized for a load resistance equal to the coil resistance. In figure 24 there is a small discrepancy between the measured and calculated power since the calculated power assumes a sinusoidal voltage but the measured voltage is not exactly sinusoidal.

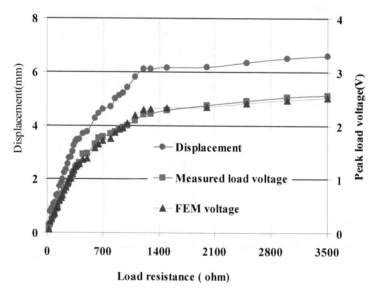

Fig. 22. Measured displacement and simulated and measured load voltage for generator A.

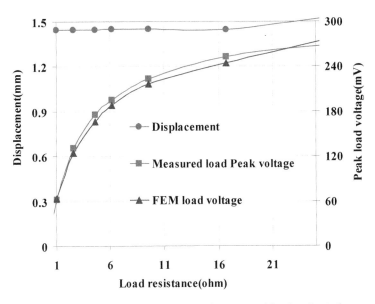

Fig. 23. Measured displacement and simulated and measured load voltage for generator B.

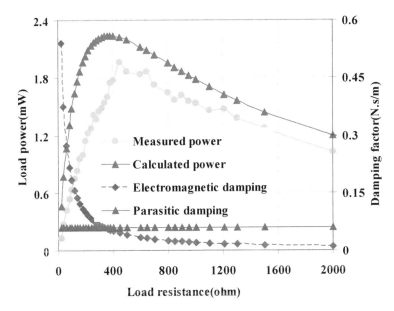

Fig. 24. Measured and calculated load power and estimated parasitic and electromagnetic damping for generator A.

In order to understand the parasitic damping effect and the optimum condition of the generator, more macro generators have been built and tested. Table 4 shows the generator parameters of macro generators C, D and E. Generators C, D, and E were tested for different acceleration levels and the vibration frequency of the shaker was swept in order to determine the resonance frequency. In generators A and B, the parasitic damping factor and the open circuit quality factor were calculated from the measured no-load displacement. However, for generators C, D and E, the no-load and load voltages at the half power bandwidth frequency were measured in order to determine the open circuit and closed circuit quality factors and hence the damping. Figures 26, 27 and 28 show the no-load voltages for different acceleration levels of generators C, D, and E, respectively. It can be seen from these figures that as the acceleration level increases, the resonance frequency shifts to a lower frequency due to the spring softening characteristic of the spring constant [31]. This indicates that the displacement of the spring constant is approaching the non-linear region. However, the resonance frequency could also shift to a higher frequency with increased acceleration level which is normally known as a spring hardening characteristic.

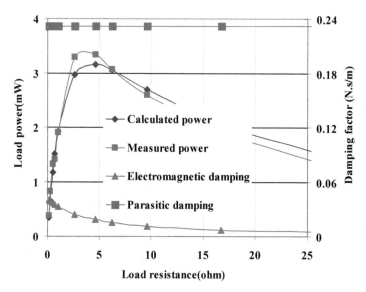

Fig. 25. Measured and calculated load power and estimated parasitic and electromagnetic damping for generator –B.

Parameters	Generator -C	Generator-D	Generator-E
Moving mass (kg)	0.019579	0.05116	0.05116
Magnet size (mm)	10 x 10 x 3	15 x 15 x 5	15 x 15 x 5
Magnet and coil gap (mm)	3.75	3.75	3.25
Coil outer diameter (mm)	19	19	28.5
Coil inner diameter (mm)	1	1	5
Coil thickness (mm)	6.5	6.5	7.5
Coil turns	1100	1100	850
Coil resistance (ohm)	46	46	18

Table 4. Generator parameters

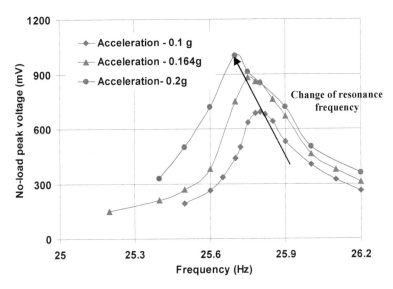

Fig. 26. No-load voltage vs frequency of generator – C.

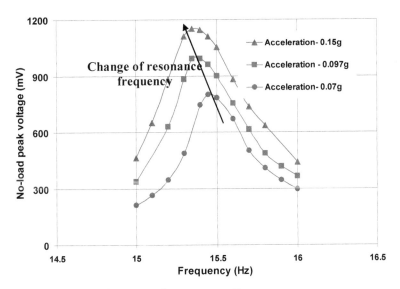

Fig. 27. No-load voltage vs frequency for generator-D.

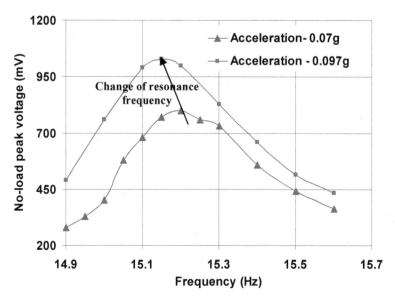

Fig. 28. No-load voltage vs frequency for generator-E.

This non-linear mass, damper and spring vibration can be defined by the standard Duffing oscillator model [31] using the following equation;

$$m\frac{d^2x}{dt^2} + D\frac{dx}{dt} + kx + k_3x^3 = F\sin\omega t \tag{31}$$

It can be seen in the above equation that a cubic nonlinear stiffness term (k_3x^3) has been added to the linear mass, damper, and spring equation. If the non–linear stiffness constant (k_3) is greater than zero then the model would represent a hardening spring constant. In this case, the resonance frequency would be shifted to the right (increase) from the linear resonance frequency with increased vibrational force, as shown in Figure 29 (a). If k_3 is less than zero, then the model would represent a softening spring constant. In this case, the linear resonance frequency would be shifted to the left (decrease) from the linear resonance frequency with increased vibrational force, as shown in Figure 29 (b).

Table 5 shows the calculated open circuit quality factor, parasitic damping factor, optimum load resistance, measured generated electrical power and load power on the optimum load resistance of each generator for different accelerations. The open circuit quality factor and the parasitic damping were calculated using the following formulas;

$$Q_{oc} = \frac{f_n}{f_2 - f_1}, \quad D_p = \frac{m\omega_n}{Q_{oc}}$$

The displacements of the generators for each load were calculated according to the equation;

$$x_{load} = \frac{F}{(D_p + D_{em})\omega_n}$$

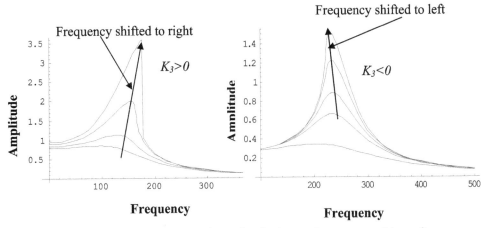

Fig. 29. Amplitude response of (a) non linear hardening spring constant (b) non linear softening spring constant.

Generator –C								
Accelerati on (g)	Q_{oc}	$X_{no\text{-}load\,(mm)}$	D_p (N.s/m)	Rc (Ω)	R_{lopt} (Ω)	D_{em} @ R_{lopt}	Generated electrical power (mW)	Max. load power (mW)
0.10	86	3.21	0.036	46	100	0.012	0.95	0.65
0.16	67	4.16	0.046	46	75	0.019	1.90	0.95
0.20	59	4.49	0.052	46	75	0.019	2.28	1.14
Generator –D								
0.07	40	2.99	0.122	46	75	0.065	1.34	0.83
0.10	36	3.76	0.133	46	75	0.065	2.23	1.38
0.15	29	4.69	0.166	46	75	0.065	3.20	1.98
Generator –E								
0.07	38	2.96	0.13	18	75	0.105	1.53	1.24
0.097	35	4.27	0.12	18	100	0.105	3.00	2.20

Table 5. Calculated parasitic damping and measured power for optimum load resistance.

The generators were also simulated using a FE transient model in order to verify the measured results with the simulated value. For example Figures 30 and 31 show the measured and simulated voltages and power respectively of the generator-C for 0.1g acceleration level. The graphs show that the measured voltages and power agree well with the FE simulation voltages. Table 5 shows that the maximum power of generator-E are transferred to the load when the parasitic damping equals the electromagnetic damping; this agrees well with the theoretical model. Generator-E delivers a maximum 81% of the generated electrical power to the load. In generator-C and D, significant electromagnetic damping is present but it is still not high enough to match the parasitic damping. The optimum load resistances for all of these generators at the maximum load power agree well with the prediction of equation (24).

It can also be seen from Table 5 that the parasitic damping is not constant for different acceleration levels. It appears that parasitic damping increases with increasing displacement. This parasitic damping depends on material properties, acceleration, size and shape of the generator structure, frequency and the vibration amplitude, as has already been explained earlier. As a consequence of this, the optimum load R_{lopt} can be different for different acceleration levels, as in the case of Generator –C.

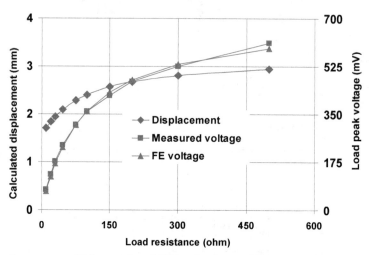

Fig. 30. Displacement and Measured and FE load voltages of generator-C for 0.1g acceleration.

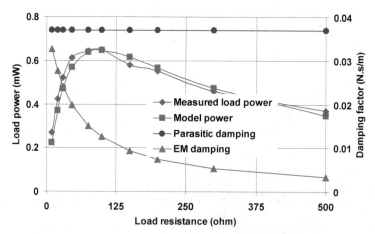

Fig. 31. Measured and calculated load power and estimated parasitic and electromagnetic damping of generator-C for 0.1g acceleration.

The power graphs for different acceleration levels of generator C & D are plotted in Figure 32 to establish the relation between the generated electrical power and the applied acceleration. According to linear theory, the generated electrical power should have a square law relation with the acceleration and an inverse square relation with total damping factor ($P_{avg} = D_{em} \dfrac{(ma)^2}{2(D_p + D_{em})^2}$). It can be seen from this graph that in practice the generated electrical power did not vary squarely with the variation of the acceleration. This is again due to the variation of parasitic damping factor, i.e as a is increased, D_p also increases and thus in practice the power has closer to a linear variation with acceleration. The next section will provide the available vibrational sources which are present in the environment since ultimate goal for energy harvester is to generate useful electrical energy from the environment.

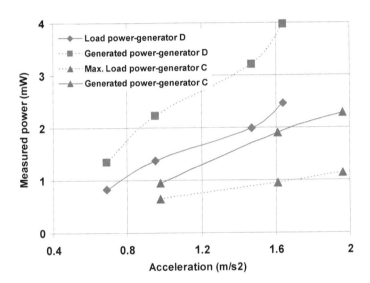

Fig. 32. Power vs acceleration

2.1 Vibrational sources

It is necessary to understand the acceleration and frequency level of different vibration sources. Since the ultimate goals of the energy harvesting device is to generate electricity from ambient sources. An overview of a variety of commonly available vibrations has already been published in several literatures [2,9]. Most of them are classified as low level vibrations which are characterised by higher frequencies and smaller amplitudes, such as industrial, automotive and structural applications and some of them are characterised by low frequency and high amplitude, such as human motions.

2.1.1 Human motion

Human motion occurs during physical activities such as walking, jogging and running. The electromagnetic vibrational generator could be mounted or attached at different locations on the human body, wired into clothes, foot-wear, a belt bag, rucksack, etc to power electronic devices using these activities. However, the amplitude, frequency, and nature of the vibration can be quite different at different locations on the human body and the acceleration would be quite high and frequencies are very low in these circumstances. For example, the acceleration level in different locations on the human body is shown in Figure 33 during walking, jogging and running on a treadmill (measured for VIBES project [31-33]). Table 6 summarises a few examples of the measured acceleration levels during walking when the accelerometer was tightly fastened on the ankle, wrist and chest. It can be seen that the maximum vertical acceleration level can be achieved at the ankle with 108 m/s^2 compared to 25 m/s^2 on the wrist and 6.6 m/s^2 on the head (front). The maximum vertical acceleration levels during walking and slow running condition were 4.9 m/s^2 (0.5g) and 9.81 m/s^2 (1g) when the accelerometer was placed in rucksack bag, as shown in Figure 34. It can be seen from this measurement that vibration is irregular and consists of high amplitude impulse like excitation rather than sinusoidal excitation and the frequency is less than 3 Hz. A resonant generator may not be the most suitable for human motion due to low frequency, high amplitude and irregular nature of human movement. Since the vibration signal in human motion tends to be non-sinusoidal random vibration, a suitable generator structure is necessary which can vibrate easily at off resonance conditions.

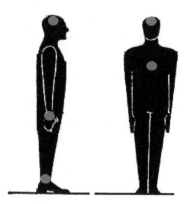

Fig. 33. Accelerometer locations on the human body.

Location	Maximum acceleration (m/s^2)
Ankle	108
Wrist	25
Chest	16
Head (front)	6.6

Table 6. Summary of acceleration levels on the human body.

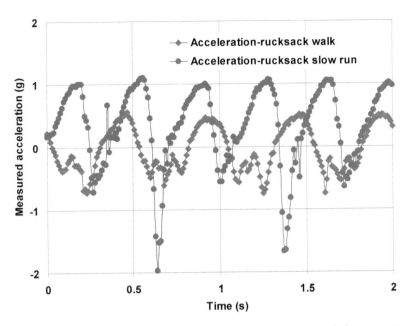

Fig. 34. Measured acceleration inside rucksack bag during walking and slow running.

2.1.2 Home appliance, machinery and automotive vibration

Vibrations from automotive applications give rise to frequencies of tens of Hz to several hundred Hz but with smaller accelerations. The vibrations generated from home appliances such as clothes, dryers, small microwave ovens and blender casings [9],[31],[32] are similar. Vibrations from rotating machines, such as pumps and fans, can include quite high frequency components, but are in general limited to relatively small accelerations. The rotational speed of these machines is constant and generates several harmonic frequency vibrations which consist of multiples of the fundamental frequency corresponding to the rotational speed. The vibration spectrum of an industrial fan (nominal speed 1500 rpm- 25 Hz), pump (nominal speed 3000 rpm-50 Hz) and air compressor unit were measured in different positions of the machines for the VIBES project [5], [33]. Figure 35, 36 and 37 show the vibration spectrum of an industrial fan and top and bottom of an air compressor unit at different positions. It can be seen from the graphs that the vibration signal is quite low amplitude with multiple vibration peak frequencies. It can be seen that all these have a peak at or near 50 Hz, 100, 150 or 250 Hz. A resonant generator structure is essential for this application in order to achieve a reasonable displacement from this very low amplitude vibration. Table 7 shows the available acceleration and frequency level of the different home appliance, machinery and automotive sources. In the following section, we present such a generator and measure the power generated from human motion when the generator is placed in a rucksack. The generator makes use of a "magnetic spring" as opposed to a mechanical spring, which could give advantages such as ease of construction, ease of tenability, and lower sensitivity to fatigue.

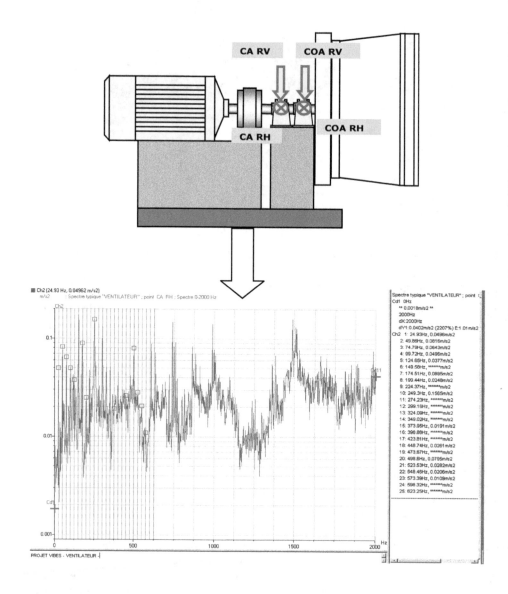

Fig. 35. Measured vibration spectrum of the industrial fan from [33].

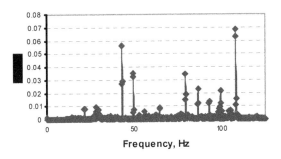

Fig. 36. Measured vibration spectrum on top of the air compressor unit [5]

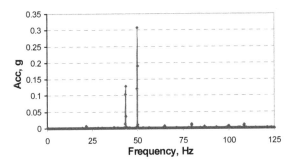

Fig. 37. Measured vibration spectrum at the bottom of an air compressor unit [5]

Vibration source	Fundamental frequency (Hz)	Acceleration (m/s²)
Car engine compartment	200	12
Base of 3-axis machine tool	70	10
Blender casing	121	6.4
Clothes dryer	121	3.5
Car instrument panel	13	3
Door frame just after door closes	125	3
Small microwave oven	121	2.5
HVAC vents in office building	60	0.2-1.5
Windows next to a busy road	100	0.7
CD on notebook computer	75	0.6
Second story floor of busy office	100	0.2
Vehicle –C (high way)	15.13	1.987
Vehicle –C (mountain)	36.88	0.0175
Vehicle-C (city)	52.87	0.0189
Industrial fan	25	0.7
Pump	50	1.4

Table 7. Home appliance, machinery and automotive vibration

2.2 Magnetic spring generator and its applications

An electromagnetic vibrational generator could be used to power electronic devices using human body activity would be considerable interest. In such an application, For example, a displacement of $x = \dfrac{ma}{D\omega_n} = 9.75$ mm could be achieved for a mass of m of 10 mg, a of 4.905 m/s^2, D of 0.4 N.s/m at f_n equal to 2 Hz.

In this case, the generated electrical power, $P = D_{em} \dfrac{(ma)^2}{2(D_p + D_{em})^2} = 1.5$ mW assuming EM damping can be made equal to parasitic damping.

It can be seen from this simple calculation that at least several cm size generators are required. In particular, a cantilever resonant generator structure would not be realistic for such a low frequency application. If we consider a 3 mm width and 50 μm thick Si or Cu cantilever beam, the length of the cantilever for a 10 mg mass and 2 Hz frequency would be:

$$k = \frac{3EI}{L^3} = m\omega_n^2 \Rightarrow L = 290 \text{ mm}$$

In order to achieve a 10 Hz frequency, a Si cantilever would have to be a 100 mm long. We present such a generator and measure the power generated from human motion when the generator is placed in a rucksack. The generator makes use of a "magnetic spring" as opposed to a mechanical spring, which could give advantages such as ease of construction, ease of tenability, and lower sensitivity to fatigue. Some of these results have already been highlighted in literature [4].

Figure 38 shows different possible configurations for the magnetic spring generator structure. The basic idea is that axially magnetized permanent magnets are placed vertically inside a tube so that facing surfaces have the same polarization. Thus, the magnets repel one another. Two magnets are fixed at both ends of the generator tube housing. A middle magnet or magnets is free to move but is suspended between both fixed end magnets in the generator housing due to the repulsive force. A coil is wrapped around the outside of the tube. When the tube is vibrated, the middle magnet vibrates up and down, and a voltage is induced in the coil. This structure can be built easily since the generator simply consists of magnets and a coil without the need for any mechanical beam. Essentially the suspended moving magnet acts like a magnetic spring constant. This construction is similar to the inductively powered torch [34], except with the addition of a magnetic spring. We know that the generation of voltage is the product of flux linkage gradient and velocity. In order to increase the flux-linkage, the single moving magnet can be replaced by two magnets separated by a soft magnetic "pole" piece, where the magnets and pole piece are glued together so that they move as a single object as shown in Figure 38 (b). The variation of flux-linkage between the single moving magnet and double moving magnets plus pole structure generator will be shown in the next section.

In order to increase the displacement, instead of using two fixed magnets, the generator could be built using only one fixed end magnet and a single moving object, as shown in Figure 38 (c). In this case, the resonant frequency would be lower and the displacement of the moving magnet would be higher compared to both fixed end magnets. The benefit of this concept in a human motion powered generator can be explained by considering the response of a spring damper system to an impulse excitation :

$$X(t) = F_0 / k, for \left[0 < t < t_0\right]$$
$$= \exp(-\xi\omega_n t) \ X \ \sin(\omega_d t + \varnothing) for \left[t > t_0\right] \tag{32}$$

When the top magnet is removed from the generator, the effective spring constant is decreased and hence the resonant frequency is decreased. Thus according to equation (32), the initial displacement will be greater and the decay rate will be slower, which would result in increased voltage and larger average power. This concept will be verified with the measured results of the real prototype which has been built and tested.

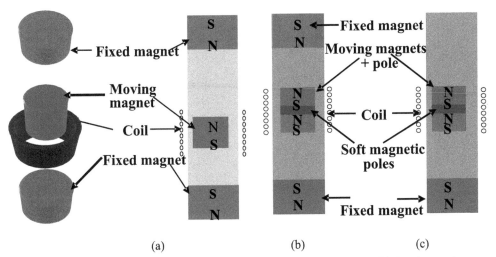

(a)   (b)   (c)

Fig. 38. Magnetic spring generator structure: (a) Single moving magnet (b) Single moving magnet replaced by two magnets + pole (c) One fixed magnet.

2.3 Analysis of generator structure

The generator structure has been modeled using Finite Element Analysis (FEA) in order to understand the spring forces which exist between the fixed and moving magnets and to understand the flux linkage with the coil. Figures 39 (a) and (b) show the results of an axi-symmetric finite element simulation of the corresponding generator structure of Figure 32 (a) and (b), respectively, showing magnetic field lines. In Figure 39 (a), a 15 x 19 mm single moving magnet is used. In Figure 39 (b), 15 x 8 mm double moving magnets and a 15 x 3 mm ferrite core are used. The overall generator dimensions are given in the next section. Figure 40 shows a plot of the radial component of the B field along a line extending from the top to the bottom of the generator for both of the generator structures. It can be seen from these field plots that the peak flux density for the double moving magnets with the pole piece is almost twice as high as for the single moving magnet generator structure. Thus, the flux gradient is higher, which translates into higher voltages and higher electromagnetic damping.

Fixed end magnets

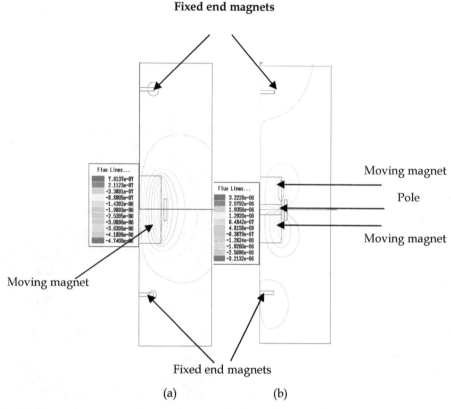

Fixed end magnets

(a)   (b)

Fig. 39. Finite element simulation, showing flux lines for a) single moving magnet b) double moving magnets plus pole generator structure.

It is also of interest to investigate the dependence of the force between the magnets poles, which can be expressed analytically [35] as:

$$F_m = \frac{\mu_0 Q_{m1} Q_{m2}}{4\pi r^2} \tag{33}$$

where $Q_m = H_c A$, H_c is the coercive force and A is the pole surface area, r is the distance between the poles. The spring constant, k, over small displacements, x, can be calculated from the linear approximation of the balanced forces equation:

$$kx = F \tag{34}$$

where the total force, F, acting on the centre magnet is given by $F = F_{m1} - F_{m2}$, F_{m1} and F_{m2} are the repulsive force magnitude on the middle magnet due to the top and bottom magnets respectively. The electromagnetic force and spring constant can be calculated from a FE transient simulation using the force vs displacement graph for the double moving magnets

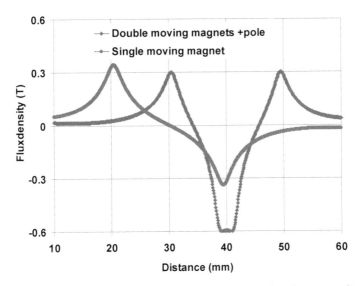

Fig. 40. Plot of radial component of flux density along a coil surface line extending from the top of the magnet tube to the bottom.

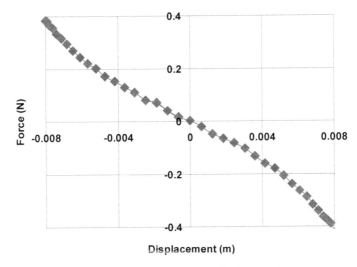

Fig. 41. Electromagnetic force vs displacement of the double moving magnets + pole generator.

plus pole structure generator which is shown in Figure 41. The resting position of the moving magnets is 4 mm away from the middle position due to the gravitational force. It can be seen from this graph that the electromagnetic force on the moving magnets is almost linear with displacement. The spring constant between the 4 mm to 8 mm region can be linearised and estimated from the graph as 61.5 N/m. In order to calculate the voltage and the electromagnetic damping factor, the flux linkage gradient is also necessary. This flux linkage gradient can be calculated from the simulated displacement and flux linkage graph as shown in Figure 42. The gradient from + 4 mm to -4 mm is 23 Wb/m. The coil can always be positioned to take advantage of this flux gradient.

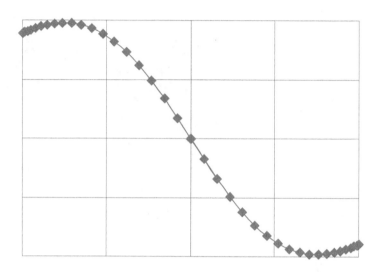

Fig. 42. FE simulated flux linkage gradient for the double moving magnets + pole generator.

2.4 Generator prototype and test results

The generator prototype consists of two opposite polarity circular magnets tightly glued to a 3 mm thick steel pole piece. This combination was inserted into a hollow Teflon tube so that it can move freely. After inserting, the two opposite polarity magnets were fixed on the both ends of the Teflon tube and 40 μm copper wire with 1000 turns coil was wrapped around the tube, offset by -4 mm away from the centre of the tube. Figure 43 shows the prototype which has been built, pictured beside a standard AA size battery. The complete dimensions and parameters of the generator are given in Table 8.

Parameters	Dimension
Tube (mm)	17 X 55
Middle magnets (mm)	15 X 8
End magnets (mm)	10 X 1
Moving mass (kg)	0.027
Coil outer diameter (mm)	18
Coil inner diameter (mm)	17
Coil thickness (mm)	6
Coil resistance (ohm)	800

Table 8. Generator parameters

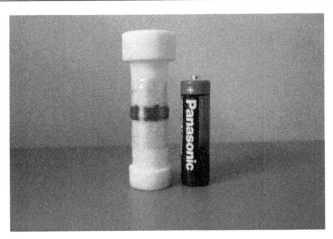

Fig. 43. Tube generator -1

2.4.1 Measured results for sinusoidal acceleration

For the first tests, the generator mounted it vertically on a force controlled electromagnetic shaker. The vibration frequency of the shaker was swept in order to determine the resonant frequency of the moving magnet combination. Any system always generates maximum vibration at the resonance condition and resonance occurs when the system natural frequency matches with the vibration frequency.

Figure 44 shows the no-load voltage vs frequency curve for 0.38259 m/s² acceleration level. It can be seen that the resonant frequency of the generator is at approximately 8 Hz. The theoretical resonant frequency, calculated from $\omega_n = \sqrt{k/m}$, where the spring constant, k, was estimated from the previous simulation, is 7.6 Hz. The measured open circuit quality factor of the generator can be estimated from the frequency response to be 18. The maximum load power measured was 14.55 µW using 7.3 kΩ load resistance where the electromagnetic damping and parasitic damping are equal. However, the aim of this generator is not to excite it with sinusoidal excitation but to excite from human movement. In the next section, we present the measured and calculated results for the prototype with human body movement.

2.4.2 Measured results of the generator for human body vibration

The generator was placed inside a rucksack and the voltage and power outputs were measured during walking and slow running conditions. An ADXL321 bi-axial accelerometer was mounted on the generator body and connected to an XR440 pocket data logger. The pocket data logger was used to measure the generator load voltage and the acceleration levels experienced by the generator.

The measured acceleration for 2 seconds data during walking and slow running conditions has already been discussed in the application section. The data shows peak acceleration levels of approximately 0.5g with a frequency of 2 Hz for walking and peak acceleration levels of approximately 1g with a frequency of approximately 2.75 Hz for slow running.

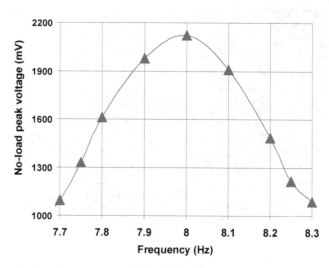

Fig. 44. Measured no-load peak voltage for half power bandwidth frequency.

Figure 45 shows the measured generated voltage graph during the walking and slow running conditions. It can be seen that the generated voltage for the slow running case is more than two times higher than for the walking case. The average load powers of the generator were measured to be 0.30 mW for walking and 1.86 mW for slow running when the coil resistance and load resistance are matched.

A second version of this generator, without a top fixed magnet, was also tested in order to compare the difference in the system frequencies, generated voltages and power levels. Figure 46 shows the measured generated voltage graph during the walking and slow running conditions. The average measured maximum load powers of the generator without top fixed magnets were 0.95 mW and 2.46 mW during the walking and slow running conditions respectively. Comparing this to the power levels obtained from the generator with both fixed magnets, it can be seen that the power level is increased three times during walking but the power level during the slow running condition is increased by only 32 %. During the slow running condition, the magnet displacement is large and it moves outside the coil where it does not generate any more voltage or power. In this case, the dimensions of the coil should be optimized for the expected displacement. In order to verify the generator voltage obtained from a single shock, the impulse responses of both generators were captured in an oscilloscope and are shown in Figure 47. It can be seen that the natural frequency and decay rate are higher in the case of the top and bottom fixed end magnet generator than the generator without top fixed magnet generator. Due to the lower spring constant the generator without the top end magnet gives a higher displacement of the moving magnet as well as a higher voltage. This is consistent with equation (32) discussed earlier.

Ultimately, the electrical energy generated will have to be stored in either a rechargeable battery or a capacitor. Thus it is of interest to investigate how much energy the generator can store over a certain period of time. The generator was placed inside a rucksack and the

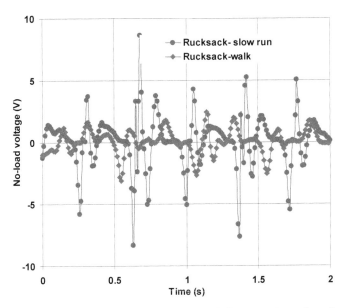

Fig. 45. Measured no-load voltage during walking and slow running when the generator was placed inside rucksack bag.

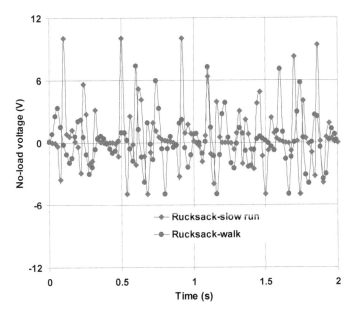

Fig. 46. Measured no-load voltage during walking and slow running for generator with only one fixed magnet.

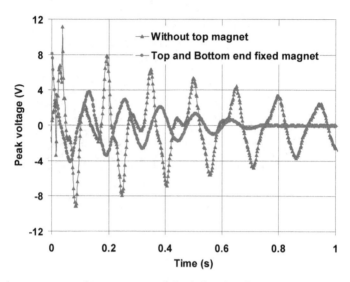

Fig. 47. Impulse responses of generators with both fixed end magnets and only one fixed end magnet.

output was connected to a rechargeable Li-MnO₂ Varta coin cell (Maxwell-ML1220) through a simple diode capacitor rectifier circuit. The battery was initially discharged to a voltage of 1.09 V. After 1 hour walking, the battery voltage had increased from 1.09 V to 1.27 V. In order to calculate the energy transfer during this time, the discharged characteristic curve of the battery was measured with a 5.5 kΩ load using the pocket logger. The measured results show that 3.54 joules of energy is transferred to the battery from the generator. This energy level is consistent with the powering of low power sensor modules without a battery. For example, the power consumption of a wearable autonomous Microsystem which consists of a light sensor, a microphone, accelerometer, microprocessor and a RF transceiver is 700 μW which is equivalent to 2.5 joules [36]. The cantilever prototypes which have been built and tested were not optimised and were not built for a specific application. Since the eventual goal would be the integration of the generator into autonomous sensors modules, a miniature generator is essential. In the following section, we will present a prototype of an optimised micro generator and its applications.

2.5 Cantilever micro generator and its applications

The theoretical model of the electromagnetic vibrational generator had been discussed and successfully verified with the different macro scale cantilever prototype generators which have been built and tested. Magnetic spring generator had also been tested using real human body applications. However all the prototypes were not optimized and it is essential to optimize the generator for specific application in order to reduce the cost and size. Few optimized micro generators had been built and tested by the University of Southampton for air compressor application [5]. The theoretical analysis and the verification of the measured results have been done as part of Author's Ph.D work for the VIBES project. The details optimization concepts of those micro generators were explained in the literature [5].

However in this section we will highlight some parts of those results. Figure 48 shows the VIBES optimized microgenerator associated with the electronics.

Fig. 48. VIBES generator associated with electronics powering the accelerometer.

The generator generates maximum power and delivers maximum power to the load when the electromagnetic damping factor equals the parasitic damping factor. Since the parasitic damping is fixed for the particular generator within linear movement, the electromagnetic damping factor can be increased by using the optimum magnet size for a particular coil, using more coil turns, and keeping the gap between the magnet and coil as low as possible. It can be seen from the following equation of electromagnetic damping that it has a square law relation with the number of coil turns and the flux linkage gradient:

$$D_{em} = \frac{N^2 (\frac{d\phi}{dx})^2}{R_c + R_l}$$

In order to achieve the maximum flux linkage gradient, the optimum magnet for a particular coil and the lowest possible gap between magnet and coil is necessary. Three optimised micro generators withs 600-, 1200- and 2300 turns coil with identical mechanical dimension have been built and tested for different acceleration levels for target application of the air compressor unit. These three generators differ only in the number of turns in the coil. Each coil was mounted on the same generator base and the same beam was assembled on each base. The 70 μm thick single crystal silicon cantilever beam which was used in the initial micro generator which was found to be too brittle to handle during assembly. The resonance frequency of the 70 μm silicon beam was 91 Hz. The targeted application is an air compressor unit which has a resonance frequency in the 50-60 Hz range. In order to reduce the resonance from 91 Hz to 60 Hz, either we have to increase the beam length or decrease the beam thickness. Instead of using a 70 μm silicon beam, a 50 μm BeCu beam was used for the optimized micro generators. BeCu has better fatigue characteristics and less brittle behaviour [5] compared to Si but both have the same Young's modulus properties. The calculated resonance frequency of the 50 μm beam is 55 Hz, which is very close to the target application frequency. This frequency can be adjusted to the required 50 Hz or 60 Hz frequency during assembly. Table 9 shows the three principle parameters of the coil.

Wire diameter, λ (μm)	No. of turns	R_{coil} (Ω)	Fill factor
25	600	100	0.67
16	1200	400	0.63
12	2300	1500	0.53

Table 9. Coil parameters.

In order to analyse the measured results with the linear modelling approach, initially the micro generator with the 600 turn coil has been tested for different acceleration levels within the linear region of the spring constant. Figure 49 shows the measured and calculated power with an optimum load resistance of 200 Ω up to 30 mg acceleration. The generated mechanical, generated electrical and the load power are calculated using the following equations:

$$ P_{mech} = \frac{1}{T}\int_0^T F \omega x_{load}\, dt = \frac{F^2}{2(D_p + D_{em})}; \quad P_{elec} = D_{em}\frac{F^2}{2(D_p + D_{em})^2}; $$

$$ P_l = D_{em}\frac{F^2}{2(D_p + D_{em})^2}\frac{R_l}{R_c + R_l}. $$

The electromagnetic damping was calculated from the FEA simulations and the parasitic damping was calculated from the measured total quality factor using the following formula:

$$ D_{em} = \frac{N^2(\frac{d\phi}{dx})^2}{R_c + R_l}, \quad D_p = \frac{m\omega_n}{Q_T} - D_{em} $$

The measured total quality factor (total damping factor) and the open circuit quality factor (parasitic damping factor) for 20mg acceleration were 119 (0.0084) and 232 (0.0041), respectively. The calculated electromagnetic damping factor was 0.0043. It can be seen from Figure 49 that the simulated power agrees with the measured power. It is also indicated from the measured quality factor that the maximum power is transferred to the load when the electromagnetic damping and the parasitic damping are equal which agrees with the theoretical model. In this generator, the measured results showed that the generated mechanical and electrical power varied with the square of the acceleration due to no variation of the parasitic damping factor. This is in contrast to the measured results for the macro scale generator, where it was found that the parasitic damping could change with the acceleration level.

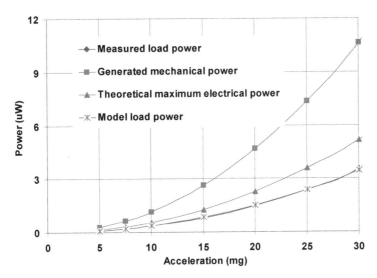

Fig. 49. Measured and calculated power of optimized micro generator for 600 turns coil.

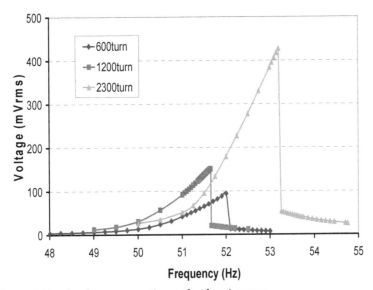

Fig. 50. Measured load voltages on optimum load resistances.

The voltages and power for 600 turn, 1200 turn and 2300 turn generators were also measured at 60 mg acceleration and with different load resistances. Figures 50 & 51 shows the measured load voltage and load power vs frequency on optimum load resistances for each generator. The optimum load resistance for 600 turns, 1200 turns and 2300 turns generators were 200Ω, 500Ω, and 4 kΩ respectively. The measured voltage shows that the generated voltages are almost proportional to the number of turns. However, the maximum load powers are constant for the three coils due to the nature of wire-wound coil technology. The measured results could not be analysed with the linear modeling approach due to the non linear behaviour of the generator for this acceleration. These non-linear results could be verified with the non-linear mass-damper- spring theory [30]. However, it is necessary to know the non linear spring constant to verify non-linear theory and this is beyond the scope of this work. Note that for practical operation it may be desirable to operate slightly away from the exact resonance frequency because of this non-linear effect.

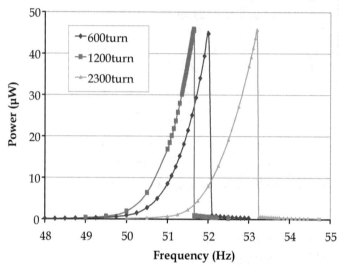

Fig. 51. Measured load powers on optimum load resistances.

2.6 Conclusions

We have presented the theory behind the modeling of the electromagnetic linear vibrational generator. The model for an electromagnetic-based vibrational energy-harvesting device has been compared to measurements on macro scale devices and FE simulation results in order to verify the modeling approach. The optimum conditions observed from measurements are shown to agree well with the model. The model and the experimental results clearly show that the generated mechanical and electrical power depends on the applied force which is the product of mass and acceleration and the damping factor: parasitic and electromagnetic damping. The parasitic damping depends on material properties, acceleration, etc., and the electromagnetic damping depends on parameters of the magnet and coil parameters. The experimental results suggest that significant parasitic damping is always present and that a maximum of 50 % of the generated electrical power could be lost in the coil's internal resistance if the parasitic damping is much greater than the electromagnetic damping. The measured results also indicate that this parasitic damping is not constant with changing acceleration level and most likely it increases with increasing the acceleration level. However in general the exact behaviour of parasitic damping with acceleration will depend on the spring material. The measured results also showed that if the magnet and coil parameters are chosen so as to maximize flux gradient and the electromagnetic damping can be made equal to the parasitic damping, then typically 80-90% of the generated electrical power can be delivered to the load. It is also clear that the generator size depends strongly on the desired system resonance frequency which in turn depends on the spring constant and the mass attached to the beam. In order to design the generator to operate at less than 10 Hz frequency, the generator should be at least a few cm size generator is size. We introduced the structure of a magnetic spring electromagnetic generator which could be beneficial over the cantilever structure for random human body activity. The prototype generators generated 0.3 - 2.46 mW when placed inside a rucksack which was worn during walking and slow running. These results indicate that a useful amount of voltage and power could be generated from the vibration of the human body. However, this device has not been optimized. The generated power of the magnetic spring generator could be increased by adding a separate coil close to the top or bottom end of the tube and connecting all coil terminals in parallel to the load. Moreover, it is possible that this generator could deliver more energy during walking if placed in other locations on the body such as the waist where higher acceleration level may be available.

3. References

[1] C. B. Williams, M.A. Harradine Shearwood, P. H. Mellor, T.S. Birch and R. B. Yates, " Development of an electromagnetic micro-generator", IEE Proc. Circuits Devices Systems, vol 148, 337-342, 2001.

[2] T. V. Buren, "Body-Worn Inertial Electromagnetic Micro-Generators" Ph.D thesis, Swiss Federal Institute of Technology Zurich.

[3] N. G. Stephen, "Energy harvesting from ambient vibration", Journal of Sound and Vibration, vol 293, 409-525, 2006.

[4] C. R. Saha, T. O'Donnell, N. Wang and P. McCloskey, "Electromagnetic generator for harvesting energy from human motion", Sensors and Actuators –A: Physical, Volume 147, Issue1, 15 September 2008.

[5] S. P. Beeby, R. N. Torah, M. J. Tudor, P. Glynne-Jones, T. O'Donnell, C.R. Saha and S. Roy "Micro electromagnetic generator for vibration energy harvesting", Journal of Micromechanics and Microengineering, 17, 1257-1265, 200

[6] C. Saha, T. O'Donnell, H. Loder, S. Beeby and J. Tudor, "Optimization of an Electromagnetic Energy Harvesting Device", IEEE Transaction on Magnetics, Volume 42, No 10, October 2006.

[7] T. O'Donnell, C. Saha, S. Beeby and J. Tudor, "Scaling Effects for Electromagnetic Vibrational power Generator", Journal of Microsystem Technology, November, 2006.

[8] P. D. Mitcheson, T. C. Green, E. M. Yeatman and A. S. Holmes, "Architectures for vibration-driven micro power generators", IEEE/ASME Journal of Microelectromechanical System, Vol 13, no. 3, 429-440, 2004.

[9] S. Roundy, P.K. Wright and J. Rabaye, "A study of low level vibrations as a power source for wireless sensor nodes", Computer Communication, vol 26, 1131-1144, 2003.

[10] P. Miao, A.S. Holmes, E.M. Yeatman and T.C. Green, "Micro-machined variable capacitors for power generation", Department of electrical and electronics engineering, Imperial college London, UK.

[11] G. Despesse, J.J. Chaillout, T. Jager, J.M. Leger, A. Vassilev, S. Basrour and B. Charlot, "High Damping Electrostatic System for Vibration Energy Scavenging", Joint sOc-EUSAI conference, Grenoble October 2005.

[12] P. Mitcheson, Stark B, P. Yeatman E, Holmes A and Green T, "Analysis and optimisation of MEMS on-chip power supply for self powering of slow moving sensors", Proc. Eurosensors XVII.

[13] F. Peano and T. Tambosso "Design and optimization of a MEMS electret-based capacitive energy scavenger" Microelectromechanical Systems, Journal of Volume 14, Issue 3, June 2005 Page(s):429 – 435.

[14] http://www.physicsclassroom.com/class/energy/u5l1c.cfm

[15] Transformer and Inductor Design Handbook, Colonel W.T. McLyman, Second Edition, Marcel Dekker Inc.New York, 1988.

[16] Electromagnetic and Electromechanical machine, Leander W. Matsch, and J. Derald Morgan, Third Edition, John Wiley and Sons.

[17] Electromechanics and Electric Machines, S. A. Nasar and L.E. Unnewehr, Second Edition, John Wiley and Sons.

[18] F. Bancel "Magnetic nodes", Journal of Physic D: Applied physics 32 (1999), 2155-2161.

[19] http://services.eng.uts.edu.au/cempe/subjects_JGZ/eet/eet_ch4.pdf

[20] John Borwick, Loudspeaker and Headphone Handbook, Focal press, Second edition

[21] P. Glynne-Jones, M. J. Tudor, S. P. Beeby, N.M. White, "An electromagnetic vibration powered generator for intelligent sensor systems", Sensors and Actuators A, 110 (2004).

[22] X. Zhang and W. C. Tang, "Viscous air damping in laterally driven microresonators", IEEE proceedings on MEMS workshop, pp. 199-204, 1994

[23] Thermoelastic damping in micro and nanomechanical systems, R Liftshitz, M Roukes, Physical Review B, Vol. 61, No. 8, 15 Feb 2000, pp.5600-5609.

[24] Internal friction in solids: 1: theory of internal friction in reeds, C Zener, Physical Review, Vol. 52, 1937, pp. 230-235.

[25] An analytical model for support loss in micromachined beam resonators, Z Hao, A Erbil, F Ayazi, Sensors and Actuators A, Vol. 109, 2003, pp. 156-164.

[26] Energy dissipation in sub-micrometer thick single crystal silicon cantilevers, J Yang, T Ono, M Esashi, J. Micromech. Systems, Vol. 11, No. 6, 2002, pp.775-783.

[27] Andrew D. Dimarogonas, Sam Haddad, Vibration for engineering, Prentice-Hall International editions, chapter 12, page 605.

[28] R. Scott Wakeland, "Use of electrodynamic drivers in thermo-acoustic refrigerators", Journal of Acoustical Society of America, 107(2), February 2000.

[29] Z. Yu, S. Backhaus and A. Jaworski, "Design and testing of a travelling wave looped tube engine for low cost electricity generators in remote and rural areas", American Institute of Aeronautics and Astronautics, 2009.

[30] Ali H. Nayfeh, Dean T. Mook, Nonlinear Oscillation, A Wiley-interscience publication.

[31] T. von Buren, P. Lukowicz and G. Troster , "Kinetic Energy Powered Computing-an Experimental Feasibility Study", Proc. 7th IEEE Int. Symposium on Wearable Computer.

[32] T. Starner and J. A. Paradiso, "Human generated power for mobile electronics", Low Power Electronics Design, CRC Press, Summer 2004.

[33] Vibrational Energy Scavenging (VIBES), FP6-IST-507911, Deliverable: Application Analysis and Specification, June 2004.

[34] A. Luzy. 1472335: magneto flash light. US Patent, October 30, 1923.

[35] S. C. Mukhopadhyay, J. Donaldson, G. Sengupta, S. Yamada, C. Chakraborty and D. Kacprzak, "Fabrication of a repulsive-type magnetic bearing using a novel arrangement of permanent magnets for vertical-rotor suspension", IEEE Transactions on Magnetics, vol. 39, No. 5, September 2003.

[36] N. B. Bharatula, S. Ossevoort, M. Stager, G. Troster, "Towards Wearable Autonomous Microsystems", Pervasive computing, proceeding of the 2nd international conference, page 225-237, Vienna, Austria, April 2004.

3

Vibration Energy Harvesting: Machinery Vibration, Human Movement and Flow Induced Vibration

Dibin Zhu
University of Southampton
UK

1. Introduction

With the development of low power electronics and energy harvesting technology, self-powered systems have become a research hotspot over the last decade. The main advantage of self-powered systems is that they require minimum maintenance which makes them to be deployed in large scale or previously inaccessible locations. Therefore, the target of energy harvesting is to power autonomous 'fit and forget' electronic systems over their lifetime. Some possible alternative energy sources include photonic energy (Norman, 2007), thermal energy (Huesgen et al., 2008) and mechanical energy (Beeby et al., 2006). Among these sources, photonic energy has already been widely used in power supplies. Solar cells provide excellent power density. However, energy harvesting using light sources restricts the working environment of electronic systems. Such systems cannot work normally in low light or dirty conditions. Thermal energy can be converted to electrical energy by the Seebeck effect while working environment for thermo-powered systems is also limited. Mechanical energy can be found in instances where thermal or photonic energy is not suitable, which makes extracting energy from mechanical energy an attractive approach for powering electronic systems. The source of mechanical energy can be a vibrating structure, a moving human body or air/water flow induced vibration. The frequency of the mechanical excitation depends on the source: less than 10Hz for human movements and typically over 30Hz for machinery vibrations (Roundy et al., 2003). In this chapter, energy harvesting from various vibration sources will be reviewed. In section 2, energy harvesting from machinery vibration will be introduced. A general model of vibration energy harvester is presented first followed by introduction of three main transduction mechanisms, i.e. electromagnetic, piezoelectric and electrostatic transducers. In addition, vibration energy harvesters with frequency tunability and wide bandwidth will be discussed. In section 3, energy harvesting from human movement will be introduced. In section 4, energy harvesting from flow induced vibration (FIV) will be discussed. Three types of such generators will be introduced, i.e. energy harvesting from vortex-induced vibration (VIV), fluttering energy harvesters and Helmholtz resonator. Conclusions will be given in section 5.

2. Energy harvesting from machinery vibration

In energy harvesting from machinery vibration, most existing devices are based on spring-mass-damping systems. As such systems are linear, these energy harvesters are also called linear energy harvesters. A generic model for linear vibration energy harvesters was first introduced by Williams & Yates (Williams & Yates, 1996) as shown in Fig. 1. The system consists of an inertial mass, m, that is connected to a housing with a spring, k, and a damper, b. The damper has two parts, one is the mechanical damping and the other is the electrical damping which represents the transduction mechanism. When an energy harvester vibrates on the vibration source, the inertial mass moves out of phase with the energy harvester's housing. There is either a relative displacement between the mass and the housing or mechanical strain.

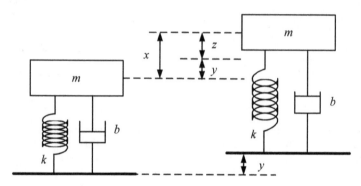

Fig. 1. Generic model of linear vibration energy harvesters

In Fig. 1, x is the absolute displacement of the inertial mass, y is the displacement of the housing and z is the relative motion of the mass with respect to the housing. Electrical energy can then be extracted via certain transduction mechanisms by exploiting either displacement or strain. The average power available for vibration energy harvester, including power delivered to electrical loads and power wasted in the mechanical damping, is (Williams & Yates, 1996):

$$P(\omega) = \frac{m\zeta_T Y^2 \left(\dfrac{\omega}{\omega_r}\right)^3 \omega^3}{\left[1 - \left(\dfrac{\omega}{\omega_r}\right)^2\right]^2 + \left[2\zeta_T \dfrac{\omega}{\omega_r}\right]^2} \qquad (1)$$

where ζ is the total damping, Y is the displacement of the housing and ω_r is the resonant frequency.

Each linear energy harvester has a fixed resonant frequency and is always designed to have a high quality (Q) factor. Therefore, a maximum output power can be achieved when the resonant frequency of the generator matches the ambient vibration frequency as:

$$P = \frac{mY^2 \omega_r^3}{4\zeta_T} \qquad (2)$$

or

$$P = \frac{ma^2}{4\zeta\omega_r} \tag{3}$$

where $a = Y\omega^2$ is the excitation acceleration. Eq. 3 shows that output power of a vibration energy harvester is proportional to mass and excitation acceleration squared and inversely proportional to its resonant frequency and damping.

When the resonant frequency of the energy harvester does not match the ambient frequency, the output power level will decrease dramatically. This drawback severely restricts the development of linear energy harvesters. To date, there are generally two possible solutions to this problem (Zhu et al., 2010a). The first is to tune the resonant frequency of a single generator periodically so that it matches the frequency of ambient vibration at all times and the second solution is to widen the bandwidth of the generator. These issues will be discussed in later sections.

There are three commonly used transduction mechanisms, i.e. electromagnetic, piezoelectric and electrostatic. Relative displacement is used in electromagnetic and electrostatic transducers while strain is exploited in piezoelectric transducer to generate electrical energy. Details of these three transducers will be presented in the next few sections.

2.1 Electromagnetic vibration energy harvesters

Electromagnetic induction is based on Faraday's Law which states that "an electrical current will be induced in any closed circuit when the magnetic flux through a surface bounded by the conductor changes". This applies whether the magnetic field changes in strength or the conductor is moved through it. In electromagnetic energy harvesters, permanent magnets are normally used to produce strong magnetic field and coils are used as the conductor. Either the permanent magnet or the coil is fixed to the frame while the other is attached to the inertial mass. In most cases, the coil is fixed while the magnet is mobile as the coil is fragile compared to the magnet and static coil can increase lifetime of the device. Ambient vibration results in the relative displacement between the magnet and the coil, which generates electrical energy. According to the Faraday's Law, the induced voltage, also known as electromotive force (e.m.f), is proportional to the strength of the magnetic field, the velocity of the relative motion and the number of turns of the coil.

Generally, there are two types of electromagnetic energy harvesters in terms of the relative displacement. In the first type as shown in Fig. 2(a), there is lateral movement between the magnet and the coil. The magnetic field cut by the coil varies with the relative movement between the magnet and the coil. In the second type as shown in Fig. 2(b), the magnet moves in and out of the coil. The magnetic field cut by the coil varies with the distance between the coil and the magnet. In contrast, the first type is more common as it is able to provide better electromagnetic coupling.

Electromagnetic energy harvesters have high output current level at the expense of low voltage. They require no external voltage source and no mechanical constraints are needed. However, output of electromagnetic energy harvesters rely largely on their size. It has been proven that performance of electromagnetic energy harvesters reduce significantly in micro scale (Beeby et al., 2007a). Furthermore, due to the use of discrete permanent magnets, it is difficult to integrate electromagnetic energy harvesters with MEMS fabrication process.

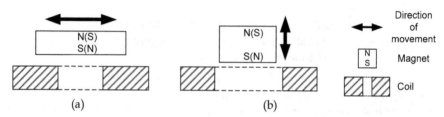

Fig. 2. Two types of electromagnetic energy harvesters

Fig. 3 compares normalized power density of some reported electromagnetic vibration energy harvesters. It is clear that power density of macro-scaled electromagnetic vibration energy harvesters is much higher than that of micro-scaled devices. This proves analytical results presented by Beeby *et al* (2007a).

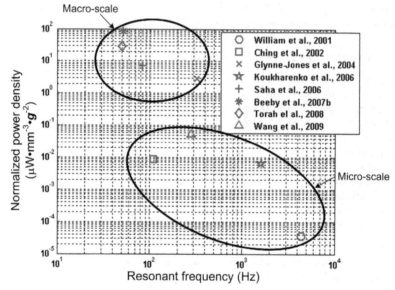

Fig. 3. Comparisons of normalized power density of some existing electromagnetic vibration energy harvesters

2.2 Piezoelectric vibration energy harvesters

The piezoelectric effect was discovered by Pierre and Jacques Curie in 1880. It is the ability of some materials (notably crystals and certain ceramics) to generate an electric potential in response to applied mechanical stress. In piezoelectric energy harvesting, ambient vibration causes structures to deform and results in mechanical stress and strain, which is converted to electrical energy because of the piezoelectric effect. The electric potential is proportional to the strain. Piezoelectric energy harvesters can work in either d_{33} mode or d_{31} mode as shown in Fig. 4. In d_{31} mode, a lateral force is applied in the direction perpendicular to the polarization direction, an example of which is a bending beam that has electrodes on its top and bottom surfaces as in Fig. 4(a). In d_{33} mode, force applied is in the same direction as the polarization direction, an example of which is a bending beam that has all electrodes on its top surfaces as in Fig. 4(b). Although piezoelectric materials in d_{31} mode normally have a lower coupling coefficients than in d_{33} mode, d_{31} mode is more commonly used (Anton and Sodano, 2007). This is because when a cantilever or a double-clamped beam (two typical structures in vibration energy harvesters) bends, more lateral stress is produced than vertical stress, which makes it easier to couple in d_{31} mode.

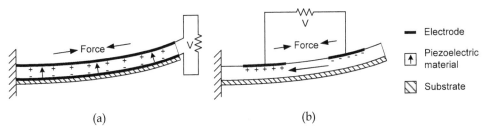

(a)   (b)

Fig. 4. Two types of piezoelectric energy harvesters (a) d_{31} mode (b) d_{33} mode

Piezoelectric energy harvesters have high output voltage but low current level. They have simple structures, which makes them compatible with MEMS. However, most piezoelectric materials have poor mechanical properties. Therefore, lifetime is a big concern for piezoelectric energy harvesters. Furthermore, piezoelectric energy harvesters normally have very high output impedance, which makes it difficult to couple with follow-on electronics efficiently. Commonly used materials for piezoelectric energy harvesting are $BaTiO_3$, PZT-5A, PZT-5H, polyvinylidene fluoride (PVDF) (Anton & Sodano, 2007). In theory, with the same dimensions, piezoelectric energy harvesters using PZT-5A has the most amount of output power (Zhu & Beeby, 2011).

Fig. 5 compares normalized power density of some reported piezoelectric vibration energy harvesters. It is found that micro-scaled piezoelectric energy harvesters have a greater power density than macro-scale device. However, due to size constraints in micro-scaled energy harvesters, the absolute amount of output power produced by the micro-scaled energy harvesters is much lower than that produced by the macro-scaled generators. Therefore, unless the piezoelectric energy harvesters are to be integrated into a micromechanical or microelectronic system, macro-scaled piezoelectric generators are preferred. Normalized power density of piezoelectric energy harvesters is about the same level as that of electromagnetic energy harvesters.

Efforts have been made to increase output power of the piezoelectric energy harvesters. Some methods include using more efficient piezoelectric materials (e.g. Macro-Fiber Composite), using different piezoelectric configurations (e.g. mode 31 or mode 33), optimizing power conditioning circuitry (Anton & Sodano, 2007), using different beam shapes (Goldschmidtboeing & Woias, 2008) and using multilayer structures (Zhu et al., 2010d).

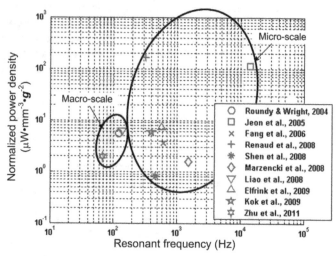

Fig. 5. Comparisons of normalized power density of some existing piezoelectric vibration energy harvesters

2.3 Electrostatic vibration energy harvesters

Electrostatic energy harvesters are based on variable capacitors. There are two sets of electrodes in the variable capacitor. One set of electrodes are fixed on the housing while the other set of electrodes are attached to the inertial mass. Mechanical vibration drives the movable electrodes to move with respect to the fixed electrodes, which changes the capacitance. The capacitance varies between maximum and minimum value. If the charge on the capacitor is constrained, charge will move from the capacitor to a storage device or to the load as the capacitance decreases. Thus, mechanical energy is converted to electrical energy. Electrostatic energy harvesters can be classified into three types as shown in Fig. 6, i.e. In-Plane Overlap which varies the overlap area between electrodes, In-Plane Gap Closing which varies the gap between electrodes and Out-of-Plane Gap which varies the gap between two large electrode plates.

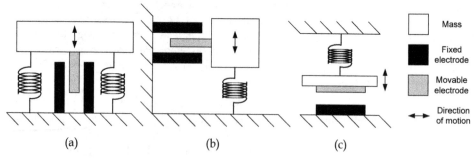

Fig. 6. Three types of electrostatic energy harvesters (a) In-Plane Overlap (b)In-Plane Gap Closing (c) Out-of-Plane Gap Closing

Electrostatic energy harvesters have high output voltage level and low output current. As they have variable capacitor structures that are commonly used in MEMS devices, it is easy to integrate electrostatic energy harvesters with MEMS fabrication process. However, mechanical constraints are needed in electrostatic energy harvesting. External voltage source or pre-charged electrets is also necessary. Furthermore, electrostatic energy harvesters also have high output impedance.

Fig. 7 compares normalized power density of some reported electrostatic vibration energy harvesters. Normalized power density of electrostatic energy harvesters is much lower than that of the other two types of vibration energy harvesters. However, dimensions of electrostatic energy harvesters are normally small which can be easily integrated into chip-level systems.

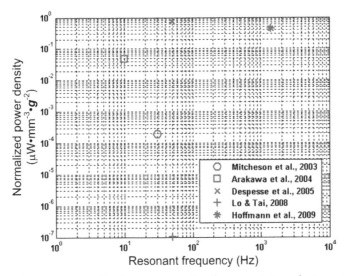

Fig. 7. Comparisons of normalized power density of some existing electrostatic vibration energy harvesters

2.4 Tunable vibration energy harvesters

As mentioned earlier, most vibration energy harvesters are linear devices. Each device has only one resonant frequency. When the ambient vibration frequency does not match the resonant frequency, output of the energy harvester can be reduced significantly. One potential method to overcome this drawback is to tune the resonant frequency of the energy harvester so that it can match the ambient vibration frequency at all time.

Resonant frequency tuning can be classified into two types. One is called continuous tuning which is defined as a tuning mechanism that is continuously applied even if the resonant frequency matches the ambient vibration frequency. The other is called intermittent tuning which is defined as a tuning mechanism that is only turned on when necessary. This tuning mechanism only consumes power during the tuning operation and uses negligible energy once the resonant frequency is matched to the ambient vibration frequency (Zhu et al., 2010a).

Resonant frequency tuning can be realized by mechanical or electrical methods. Realizations of mechanical tuning include changing the dimensions of the structure, moving the centre of gravity of proof mass and changing spring stiffness continuously or intermittently. Most mechanical tuning methods are efficient in frequency tuning and suitable for in situ tuning, i.e. tuning the frequency while the generator is in operation. However, extra systems and energy are required to realize the tuning. Electrical methods typically adjust electrical loads of the generator to tune the resonant frequency. This is much easier to implement. Closed-loop control is necessary for both mechanical tuning and electrical tuning so that the resonant frequency can match the vibration frequency at all times. As most of the existing vibration energy harvesters are based on cantilever structures, only frequency tuning of cantilever structures will be discussed in this section.

2.4.1 Variable dimensions

The spring constant of a resonator depends on its materials and dimensions. For a cantilever with a mass at the free end, the resonant frequency, f_r, is given by (Blevins, 2001):

$$f_r = \frac{1}{2\pi} \sqrt{\frac{Ywh^3}{4l^3 (m + 0.24m_c)}} \tag{4}$$

where Y is Young's modulus of the cantilever material; w, h and l are the width, thickness and length of the cantilever, respectively. m is the inertial mass and m_c is the mass of the cantilever. The resonant frequency can be tuned by adjusting all these parameters. However, it is difficult to change the width and thickness of a cantilever in practice. Only changing the length is feasible. Furthermore, modifying length is suitable for intermittent tuning. The approach requires an extra clamper besides the cantilever base clamp. This extra clamper can be released and re-clamped in different locations for various resonant frequencies. There is no power required to maintain the new resonant frequency. This approach has been patented (Gieras et al., 2007). However, due to its complexity, there is few research reported on this method.

2.4.2 Variable centre of gravity of the inertial mass

The resonant frequency can be adjusted by moving the centre gravity of the inertial mass. The ratio of the tuned frequency, f_r', to the original frequency, f_r, is (Roylance & Angell, 1979):

$$\frac{f_r'}{f_r} = \sqrt{\frac{1}{3} \cdot \frac{r^2 + 6r + 2}{8r^4 + 14r^3 + \frac{21}{2}r^2 + \frac{2}{3}}} \tag{5}$$

where r is the ratio of the distance between the centre of gravity and the end of the cantilever to the length of the cantilever.

This approach was realized and reported by Wu et al (2008). The tunable energy harvester consists of a piezoelectric cantilever with two inertial masses at the free end. One mass was fixed to the cantilever while the other part can move with respect to the fixed mass. Centre of gravity of the inertial mass could be adjusted by changing the position of the movable mass. The resonant frequency of the device was successfully tuned between 180Hz and 130Hz. The output voltage dropped with increasing resonant frequency.

2.4.3 Variable spring stiffness

Another method to tune the resonant frequency is to apply an external force to change stiffness of the spring. This tuning force can be electrostatic, piezoelectric, magnetic or other mechanical forces. However, electrostatic force requires very high voltage. In addition, spring stiffness can also be changed by thermal expansion but energy consumption in this method is too high compared to power generated by vibration energy harvesters. Therefore, these two methods are not suitable for frequency tuning in vibration energy harvesting. In this section, only frequency tuning by piezoelectric, magnetic and direct forces is discussed.

Peters et al (2008) reported a tunable resonator suitable for vibration energy harvesting. The resonant frequency tuning was realised by applying a force using piezoelectric actuators. A piezoelectric actuator was used because piezoelectric materials can generate large forces with low power consumption. The tuning voltage was chosen to be ±5V resulted in a measured resonance shift of ±15% around the initial resonant frequency of 78 Hz, i.e. the tuning range was from 66Hz to 89Hz. A closed-loop phase-shift control system was later developed to achieve autonomous frequency tuning (Peters et al., 2009). Eichorn et al (2010) presented a piezoelectric energy harvester with a self-tuning mechanism. The tuning system contains a piezoelectric actuator to provide tuning force. The device has a tuning range between 188Hz and 150Hz with actuator voltage from 2V to 50V. These are two examples of continuous tuning.

An example of applying magnetic force to tune the resonant frequency was reported by Zhu et al (2010b) who designed a tunable electromagnetic vibration energy harvester. Frequency tuning was realised by applying an axial tensile magnetic force to a cantilever structure as shown in Fig. 8.

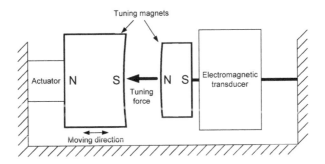

Fig. 8. Frequency tuning by applying magnetic force (reproduced from (Zhu et al., 2010b))

The tuning force was provided by the attractive force between two tuning magnets with opposite poles facing each other. One magnet was fixed at the free end of a cantilever while the other was attached to an actuator and placed axially in line with the cantilever. The distance between the two tuning magnets was adjusted by the linear actuator. Thus, the axial load on the cantilever, and hence the resonant frequency, was changed. The areas where the two magnets face each other were curved to maintain a constant gap between them over the amplitude range of the generator. The tuning range was from 67.6 to 98Hz by changing the distance between two tuning magnets from 5 to 1.2mm. The tuning mechanism does not affect the damping of the micro-generator over most of the tuning range. However, when the tuning force became larger than the inertial force caused by vibration, total damping increased and the output power was less than expected from theory. A control system was designed for this energy harvester (Ayala-Garcia et al., 2009). Energy consumed in resonant frequency tuning was provided by the energy harvester itself. This is the first reported autonomous tunable vibration energy harvester that operates exclusively on the energy harvester.

Resonant frequency of a vibration energy harvester can also be tuned by applying a direct mechanical force (Leland and Wright, 2006). The energy harvester consisted of a double clamped beam with a mass in the centre. The tuning force was compressive and was applied using a micrometer at one end of the beam. The tuning range was from 200 to 250 Hz. It was determined that a compressive axial force could reduce the resonance frequency of a vibration energy harvester, but it also increased the total damping. The above two devices are examples of intermittent tuning.

2.4.4 Variable electrical loads

All frequency tuning methods mentioned above are mechanical methods. Mechanical methods generally have large tuning range. However, they require a load of energy to realise. This is crucial to vibration energy harvesting where energy generated is quite limited. Therefore, electrical tuning method is introduced. The basic principle of electrical tuning is to change the electrical damping by adjusting electrical loads, which causes the power spectrum of the generator to shift.

Charnegie (2007) presented a piezoelectric energy harvester based on a bimorph structure and adjusted its resonant frequency by varying its load capacitance. The test results showed that if one piezoelectric layer was used for frequency tuning while the other one was used for energy harvesting, the resonant frequency can be tuned an average of 4 Hz with respect to the original frequency of 350 Hz by adjusting the load capacitance from 0 to 10 mF. If both layers were used for frequency tuning, the tuning range was an average of 6.5 Hz by adjusting the same amount of load capacitance. However, output power was reduced if both layers were used for frequency tuning while if only one layer was used for frequency tuning, output power remained unchanged.

Another electrically tunable energy harvester was reported by Cammarano *et al* (2010). The resonant frequency of the electromagnetic energy harvester was tuned by adjusting electrical loads, i.e. resistive, capacitive and inductive loads. The tuning range is between 57.4 and 66.5Hz. However, output power varied with changes of electrical loads.

2.5 Vibration energy harvesters with wide bandwidth

The other solution to increase the operational frequency range of a vibration energy harvester is to widen its bandwidth. Most common methods to widen the bandwidth include using a generator array, using nonlinear and bi-stable structures. In this section, details of these approaches will be covered.

2.5.1 Generator array

A generator array consists of multiple small energy harvesters, each of which has different dimensions and masses and hence different resonant frequencies. Thus, the assembled array has a wide operational frequency range whilst the Q-factor does not decrease. The overall power spectrum of a generator array is a combination of the power spectra of each small generator as shown in Fig. 9. The frequency band of the generator is thus essentially increased. The drawback of this approach is the added complexity in design and fabrication of such array and the increased total volume of the device depending upon the number of devices in the array.

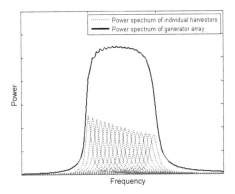

Fig. 9. Frequency spectrum of a generator array

Sari *et al* (2008) reported a micromachined electromagnetic generator array with a wide bandwidth. The generator consisted of a series of cantilevers with various lengths and hence resonant frequencies. Cantilevers were carefully designed so that they had overlapping frequency spectra with the peak powers at similar but different frequencies. This resulted in a widened bandwidth as well as an increase in the overall output power. Coils were printed on cantilevers while a large magnet was fixed in the middle of the cantilever array. Experimentally, operational frequency range of this device is between 3.3 and 3.6 kHz where continuous power of 0.5µW was generated.

A multifrequency piezoelectric generator intended for powering autonomous sensors from background vibrations was presented by Ferrari *et al* (2008). The generator consisted of three bimorph cantilevers with different masses and thus natural frequencies. Rectified outputs were fed to a single storage capacitor. The generator was used to power a batteryless sensor module that intermittently read the signal from a passive sensor and sent the measurement information via RF transmission, forming an autonomous sensor system. Experimentally, none of the cantilevers used alone was able to provide enough energy to operate the sensor module at resonance while the generator array was able to power the sensor node within wideband frequency vibrations.

2.5.2 Nonlinear structures

The theory of vibration energy harvesting using nonlinear generators was investigated by Ramlan (2009). Numerical and analytical showed that bandwidth of the nonlinear system depends on the damping ratio, the nonlinearity and the input acceleration. Ideally, the maximum amount of power harvested by a nonlinear system is the same as the maximum power harvested by a linear system. There are two types of nonlinearity, i.e. hard nonlinearity and soft nonlinearity as shown in Fig. 10. It is worth mentioning that output power and bandwidth depend on the approaching direction of the vibration frequency to the resonant frequency. For a hard nonlinearity, this approach will only produce an improvement when approaching the device resonant frequency from a lower frequency. For a soft nonlinearity, this approach will only produce an improvement when approaching the device resonant frequency from a higher frequency. It is unlikely that these conditions can be guaranteed in real application, which makes this method very application dependent.

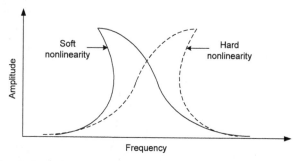

Fig. 10. Soft and hard Nonlinearity

Most reported nonlinear vibration energy harvester is realized by using a magnetic spring. Burrows *et al* (2007, 2008) reported a nonlinear energy harvester consisting of a cantilever spring with the non-linearity caused by the addition of magnetic reluctance forces. The device had a flux concentrator which guided the magnetic flux through the coil. The reluctance force between the magnets and the flux concentrator resulted in non-linearity. It was found experimentally that the harvester had a wider bandwidth during an up-sweep, i.e. when the excitation frequency was gradually increased while the bandwidth was much narrower during a down-sweep, i.e. when the excitation frequency was gradually decreased. This is an example of hard nonlinearity.

Another example of nonlinear vibration energy harvester is a tunable electromagnetic vibration energy harvester with a magnetic spring, which combined a manual tuning mechanism with the non-linear structure (Spreemann et al., 2006). This device had a rotary suspension and magnets as nonlinear springs. It was found in the test that the bandwidth of the device increased as magnetic force became larger, i.e. non-linearity increased.

A numerical analysis of nonlinear vibration energy harvesters was recently reported (Nguyen & Halvorsen, 2010). Analytical results showed that soft nonlinear energy harvesters have better performance than hard nonlinear energy harvesters. This is yet to be verified by experiments.

2.5.3 Bi-stable structures

Ramlan (2009) also studied bi-stable structures for energy harvesting (also termed the snap-through mechanism). Analysis revealed that the amount of power harvested by a bistable device is $4/\pi$ greater than that by the tuned linear device as the device produces a squarewave output for a given sinusoidal input. Numerical results also showed that more power is harvested by the mechanism if the excitation frequency is much less than the resonant frequency. Bi-stable devices also have the potential to cope with the mismatch between the resonant frequency and the vibration frequency.

Ferrari *et al* (2009) reported a nonlinear generator that exploits stochastic resonance with white-noise excitation. A piezoelectric beam converter was coupled to permanent magnets creating a bi-stable system bouncing between two stable states in response to random excitation. Under proper conditions, this significantly improved energy harvesting from wide-spectrum vibrations. The generator was realized by screen printing low-curing-temperature lead zirconate titanate (PZT) films on steel cantilevers and excited with white-noise vibrations. Experimental results showed that the performances of the converter in terms of output voltage at parity of mechanical excitation were markedly improved.

Mann *et al* (2010) investigated a nonlinear energy harvester that used magnetic interactions to create an inertial generator with a bistable potential well. The motivating hypothesis for this work was that nonlinear behavior could be used to improve the performance of an energy harvester by broadening its frequency response. Theoretical investigations studied the harvester's response when directly powering an electrical load. Both theoretical and experimental tests showed that the potential well escape phenomenon can be used to broaden the frequency response of an energy harvester.

Erturk *et al* (2009) introduced a piezomagnetoelastic device for substantial enhancement of piezoelectric vibration energy harvesting. Electromechanical equations describing the nonlinear system were given along with theoretical simulations. Experimental performance of the piezomagnetoelastic generator exhibited qualitative agreement with the theory, yielding large-amplitude periodic oscillations for excitations over a frequency range. Comparisons were presented against the conventional case without magnetic buckling and superiority of the piezomagnetoelastic structure as a broadband electric generator was proven. The piezomagnetoelastic generator resulted in a 200% increase in the open-circuit voltage amplitude (hence promising an 800% increase in the power amplitude).

2.6 Summary

Eq. 3 gives a good guideline in designing vibration energy harvester. The maximum power converted from the mechanical domain to the electrical domain is proportional to the mass and vibration acceleration squared and inversely proportional to the resonant frequency as well as total damping. This means that more power can be extracted if the inertial mass is increased or energy harvesters can work in the environment where the vibration level is high. For a fixed resonant frequency, the generator has to be designed to make the mechanical damping as low as possible. For an energy harvester with constant damping, the generated electrical power drops with an increase of the resonant frequency.

However, as vibration energy harvesters are usually designed to have a high Q-factor for better performance, the generated power drops dramatically if resonant frequencies and ambient vibration frequencies do not match. Therefore, most reported generators are designed to work only at one particular frequency. For applications such as moving vehicles, human movement and wind induced vibration where the frequency of ambient vibration changes periodically, the efficiency of energy harvesters with one fixed resonant frequency is significantly reduced since the generator will not always be at resonance. This drawback must be overcome if vibration energy harvesters are to be widely applicable in powering wireless systems.

Tuning the resonant frequency of a vibration energy harvester is a possible way to increase its operational frequency range. It requires a certain mechanism to periodically adjust the resonant frequency so that it matches the frequency of ambient vibration at all times.

The suitability of different tuning approaches will depend upon the application, but in general terms the key factors for evaluating a tuning mechanism for adjusting the resonant frequency of vibration energy harvesters are as follows. First, energy consumed by the tuning mechanism must not exceed the energy generated. Second, tuning range should be large enough for certain applications. Third, tuning mechanism should achieve a suitable degree of frequency resolution. Last but not least, tuning mechanism should have as little effect on total damping as possible. Furthermore, intermittent tuning is preferred over continuous tuning as it is only on when necessary and thus saves energy.

It is important to mention that efficiency of mechanical tuning methods depends largely on the size of the structure. The smaller the resonator, the higher the efficiency of the tuning mechanism. Efficiency of resonant frequency tuning by adjusting the electrical load depends on electromechanical coupling. The better the coupling, the larger the tuning range. Mechanical tuning methods normally provide large tuning range compared to electrical tuning methods while electrical tuning methods require less energy than mechanical tuning methods.

Operational frequency range of a vibration energy harvester can be effectively widened by designing an energy harvester array consisting of multiple small generators which work at various frequencies. Thus, the assembled energy harvester has a wide operational frequency range whilst the Q-factor does not decrease. However, this array must be designed carefully so that individual harvesters do not affect each other, which makes it more complex to design and fabricate. In addition, only a portion of individual harvesters contribute to power output at a particular source frequency. Therefore, this approach is not volume efficient. Furthermore, non-linear energy harvesters and harvesters with bi-stable structures are another two solutions to increase the operational frequency range of vibration energy harvesters. They can improve performance of the generator at higher and lower frequency bands relative to its resonant frequency, respectively. However, the mathematical modelling of these energy harvesters is much more complicated than that of linear generators, which increases the complexity in design and implementation. In addition, there is hysteresis in non-linear energy harvesters. Performance during down-sweep (or up-sweep) can be worse than that during up-sweep (or down-sweep) or worse than the linear region depending on sweep direction. Therefore, when designing nonlinear energy harvesters, this must be taken into consideration. In contrast, energy harvesters with bi-stable structures are less frequency dependent, which makes it a potentially better solution.

In summary, some most practical methods to increase the operation frequency range for vibration energy harvesting include:

- changing spring stiffness intermittently (preferred) or continuously;
- adjusting electrical loads;
- using generator arrays;
- employing non-linear and bi-stable structures.

3. Energy harvesting from human movement

The human body contains huge amount of energy. The kinetic energy from human movement can be harvested and converted to electrical energy. The electrical energy produced can be used to power other wearable electronics, for example, a watch and a heart rate monitor. It can also be used to charge portable electronics, such as mobile phones, mp3 players or even laptops. Researches have been done to study movement of different parts of a human body. It was found that upper human body produces movement with frequencies less than 10Hz while frequencies of movement from lower human body are between 10 and 30Hz (von Buren, 2006). The first prototype of the electronic device powered by human movement is an electronic watch developed by SEIKO in 1986. Two years later, SEIKO launched the world's first commercially available watch, called AGS. Since then, more and more human-powered electronic devices have come to the market and researches in this area have drawn more attention (Romero et al., 2009). So far, two common types of human energy harvesters are energy harvesting shoes and backpacks.

3.1 Shoes

Energy harvesters in shoes are based on either pressure of the human body on the shoe sole or the kicking force during walking.

Kymissis *et al* (1998) studied energy harvesters mounted on sneakers that generated electrical energy from the pressure on the shoe sole. Output power of three types of energy harvesters was reported. The first energy harvesters had multilayer laminates of PVDF, the second one contained a PZT unimorph and the third one was a rotary electromagnetic generator. The PVDF and PZT elements were mounted between the removable insole and rubber sole. The PVDF stack was in the front of the shoe while the PZT unimorph was at the heel. The electromagnetic generator was installed under the heel. Experimentally, the three generators produced average power of 1.8mW, 1.1mW and 230mW, respectively.

Carroll and Duffy (2005) reported a sliding electromagnet generator placed inside the shoe sole for energy harvesting. This device extracted electrical energy from the kicking force during walking. The generator consists of a set of three coils with magnets moving inside the coils. Experimentally, this generator produced up to 8.5mW of power at 5Hz. A smaller set of three generators was also presented. This set delivered up to 230μW of power at 5Hz.

3.2 Backpacks

There are also two types of energy harvesting from backpacks. One utilises linear vertical movement of the backpacks to generate electrical energy and the other is based on stress on the strips of the backpacks.

Rome *et al* (2005) studied a backpack that converted kinetic energy from the vertical movement of a backpack to electrical energy. The backpack consisted of a linear bearing and a set of springs suspended the load relative to a frame and shoulder harness. The load could move vertically relative to the frame. This relative motion was then converted to electrical energy using a rotary electric generator with a rack and pinion. This system was demonstrated to generate a maximum power of approximately 7.37W. Although the backpack does generate significant power levels, the additional degree of freedom provided to the load could impair the user's dexterity and lead to increased fatigue.

Saha *et al* (2008) reported a nonlinear energy harvester with guided magnetic spring for energy harvesting from human movement. The average measured maximum load powers of the generator without top fixed magnets were 0.95mW and 2.46mW during walking and slow running condition, respectively.

Energy harvesting from a backpack with piezoelectric strips was reported by Granstrom *et al* (2007). The traditional strap of the backpack was replaced by one made of PVDF. PVDF was chosen due to its high flexibility and strength. In the test, a preload of around 40N was applied to the straps to simulate the static weight in the backpack while a 20N sine wave with a frequency of 5Hz was applied to simulate the alternating load in the backpack. Strips with PVDF of 28μm and 52μm were compared. Maximum power generated in these two strips was 3.75mW and 1.36mW, respectively.

Another backpack targeted straps as locations for piezoelectric generators was reported by Feenstra *et al* (2008). A piezoelectric stack was placed in series with the backpack straps. The tension force that the piezoelectric stack receives from the cyclic loading is mechanically amplified and converted into a compressive load. The average power output measured when walking on a treadmill with a 40lb load was reported as 176μW. The maximum power output for the device was expected to be 400μW.

3.3 Summary

Energy harvesting from human movement is quite different from energy harvesting from machinery vibration due to some special characters. First, human movement has low frequency (<30Hz) and large displacement (several mm or cm). Second, human movement is not sinusoidal. It is normally random. Therefore, resonant energy harvesters that are widely used in energy harvesting from machinery are not suitable for this application. Last but not least, energy harvesters to be worn on human body should have reasonable size and weight so that they will not affect normal human activity. Table 1 summarizes some reported energy harvesters from human movement.

4. Energy harvesting from flow induced vibrations

The turbine generator is the most mature method for flow energy harvesting. However, the efficiency of conventional turbines reduces with their sizes due to the increased effect of friction losses in the bearings and the reduced surface area of the blades. Furthermore, rotating components such as bearings suffer from fatigue and wear, especially when miniaturised. These drawbacks of turbine generators urges emergence of a new area in energy harvesting, i.e. energy harvesting from flow induced vibration. The flow here

Generator type	Position	Operational principle	Output power (mW)
PVDF laminates	front of the shoe	Pressure	1.8
PZT unimorph	heel		1.1
electromagnetic	heel		230
electromagnetic	heel	Kicking force	8.5
nonlinear	backpack	Walking	0.95
		Running	2.46
PVDF strip		Preload: 40N 20N sine wave@5Hz	3.75
			1.36
Piezoelectric stack		Walking	0.176

Table 1. Comparisons of some existing energy harvesters from human movement

includes both liquid flow and air flow. There are three main types of energy harvester of this kind. They are energy harvesting from vortex-induced vibration (VIV), flutter energy harvesters and energy harvesters with Helmholtz resonators. Principles and reported devices will be presented in this section.

4.1 Energy harvesting from vortex-induced vibrations

Flow-induced vibration, as a discipline, is very important in our daily life, especially in civil engineering. Generally, scientists try to avoid flow-induced vibration in buildings and structures to reduce possible damage. Recently, such vibration has been investigated as an energy source that can be used to generate electrical energy. Two types of flow-induced vibration are studied so far: vortex-induced vibration and flutter.

4.1.1 Principles

When a fluid flows toward the leading edge of a bluff body, the pressure in the fluid rises from the free steam pressure to the stagnation pressure. When the flow speed is low, i.e. the Reynolds number is low, pressure on both sides of the bluff body remains symmetric and no turbulence appears. When the flow speed is increased to a critical value, pressure on both sides of the bluff body becomes unstable, which causes a regular pattern of vortices, called vortex street or Kármán vortex street as shown in Fig. 11. Certain transduction mechanisms can be employed where vortices happen and thus energy can be extracted. Sanchez-Sanz *et al* (2009) studied the feasibility of energy harvesting based on the Kármán vortex street and proposed several design rules of such micro-resonator. This method is suitable both air flow and liquid flow.

Flutter is a self-feeding vibration where aerodynamic forces on an object couple with a structure's natural mode of vibration to produce rapid periodic motion. Flutter can occur in any object within a strong fluid flow, under the conditions that a positive feedback occurs between the structure's natural vibration and the aerodynamic forces. Flutter can be very disastrous. The worst example of flutter is the disaster of Tacoma Narrows Bridge that collapsed due to the aeroelastic flutter. However, such vibrant movement makes it an ideal source for energy harvesting. This method is normally only suitable for air flow as damping in liquid flow is very high, which makes flutter less likely to happen.

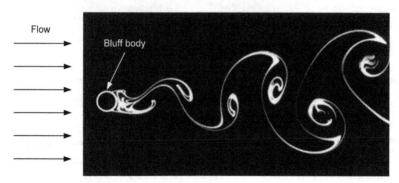

Fig. 11. An example of Kármán vortex street

4.1.2 Energy harvesting in liquid flow

The most famous energy harvester based on Kármán vortex street is the 'Energy Harvesting Eel' (Allen & Smits, 2001; Taylor et al., 2001). Fig. 12 shows a schematic of the device. The 'eel' was a flexible membrane with PVDF on it. It is riveted a certain distance away behind a fixed bluff body. The vortices behind the bluff body caused the 'eel' to swing from one end to the other. Electrical energy can then be generated by the PVDF from such movement. However, no detailed test results were reported.

Fig. 12. Schematic of the 'Energy Harvesting Eel' (top view)

Wang and Pham (2011a) reported a small scale water flow energy harvester based on Kármán vortex street. The energy harvester had a flexible diaphragm on which a piezoelectric film (PVDF) was attached. There was a chamber below the diaphragm where the water flows. A bluff body iwas placed at the centre of the chamber. When the water flew past the bluff body, vortex street occurred. The diaphragm moved up and down with the vortices. The movement of the diaphragm bent the piezoelectric film and thus generated electrical energy. Experimental results showed that an open circuit output voltage of $0.12V_{pp}$ and an instantaneous output power of 0.7nW were generated when the pressure oscillated with amplitude of 0.3kPa and a frequency of 52Hz. Its active volume was 50mm × 26mm × 15mm. The active volume is defined as the product of the area of the diaphragm times the thickness of the device.

Similar devices without the bluff body were also studied by Wang *et al* (2010a, 2010b, 2011b). Both piezoelectric and electromagnetic transducers were used. Table 2 lists their test results.

Transducer	Output power (µW)	Open circuit voltage (V)	Flow pressure (Pa)	Flow frequency (Hz)	Active volume (mm × mm × mm)
Electromagnetic (Wang, 2010a)	0.4	0.01	254	30	900 × 600 × 400
Piezoelectric (Wang, 2011b)	$0.45×10^{-3}$	0.072	20.8k	45	23 × 15 × 10
Piezoelectric (Wang, 2010b)	0.2	2.2	1196	26	50 × 30 × 7

Table 2. Comparison of Wang's work

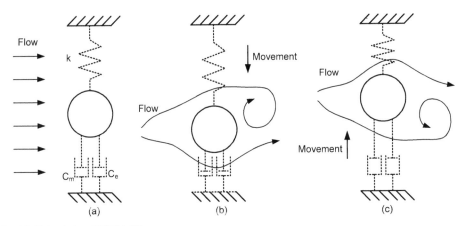

Fig. 13. Principle of VIVACE

Another type of energy harvesters in water based on Kármán vortex street is called Vortex Induced Vibration for Aquatic Clean Energy (VIVACE) (Bernitsas, 2006). The principle of this energy harvester is slightly different from that of the ones mentioned above. Instead of using the vortices created by a fixed bluff body, this energy harvester uses movement of the bluff body caused by the vortices it produces itself to generate power. When a flow passes a mobile bluff body, vortices are formed. The formation of a vortex alternately above and below the cylindrical bluff body forces an alternating vertical motion of the cylinder, the energy of which can be extracted (as shown in Fig. 13.). Note that the bluff body was designed to be restricted to have only one degree of freedom. Electromagnetic transducer was used to generator electrical energy. Multiple cylinders can be used to form arrays depending on applications.

Such devices are currently available only in large scales. Six different scales of VIVACE with power lever between 50kW and 1GW were reported so far. More work needs to be done to minimize it so that it can be used to power wireless sensor nodes. Barrero-Gil *et al* (2010) published a model for such energy harvesting method. Several design rules were summarized. Furthermore, the authors concluded that it is fairly straightforward to minimize such devices.

4.1.3 Energy harvesting in airflow

One method of energy harvesting based on Kármán vortex street, called flapping-leaf, has been reported by Li and Lipson (2011). The flapping-leaf energy harvester had the same principle as the 'energy harvesting eel' while it was only designed to work in airflow. The device consisted of a PVDF cantilever with one end clamped on a bluff body and the other end connected to a triangular plastic leaf. When the airflow passed the bluff body, the vortices produced fluctuated the leaf and thus the PVDF cantilever to produce electrical energy. The energy harvester generated a maximum output power of 17µW under the wind of 6.5m·s^{-1}. Dimensions of the PVDF cantilever was 73mm × 16mm × 40µm.

Dunnmon *et al* (2011) reported a piezoelectric aeroelastic energy harvester. It consists of a flexible plate with piezoelectric laminates which was placed behind a bluff body. It was excited by a uniform axial flow field in a manner analogous to a flapping flag such that the system delivered power to an electrical impedance load. In this case, the bluff body was in the shape of a standard NACA 0015 rather than a cylinder. The beam was made of 2024-T6 aluminium and an off-the-shelf piezoelectric patch was mounted close to the clamped end of the beam in the centre along the width of the beam. Experimental results showed that a RMS output power of 2.5mW can be derived under a wind of 27m·s^{-1}. The generator was estimated to have an efficiency of 17%. The plate had dimensions of 310mm × 101mm × 0.39mm and the bluff body has a length of 550mm. Dimensions of the piezoelectric laminate were 25.4mm × 20.3mm × 0.25mm.

Jung and Lee (2011) recently presented a similar electromagnetic energy harvester as VIVACE. Instead of operating under water, this device was designed to work under air flow. In addition, this device had a fixed cylinder bluff body in front of the mobile cylinder. These two cylinders had the same dimensions. It was found that the displacement of the mobile cylinder largely depends on the distance between the two cylinders and the maximum displacement can be achieved when this distance was between three and six times of the cylinder diameter. In the experiments, a prototype device can produce an average output power of 50-370mW under wind of 2.5-4.5 m·s^{-1}. Both cylinders had a diameter of 5cm and a length of 0.85m.

Zhu *et al* (2010c) presented a novel miniature wind generator for wireless sensing applications. The generator consisted of a wing that was attached to a cantilever spring made of beryllium copper. The airflow over the wing caused the cantilever to bend upwards, the degree of bending being a function of the lift force from the wing and the spring constant. As the cantilever deflects downwards, the flow of air is reduced by the bluff body and the lift force reduced causing the cantilever to spring back upwards. This exposes it to the full airflow again and the cycle is repeated (as shown in Fig. 14). When the frequency of this movement approaches the resonant frequency of the structure, the wing has the maximum displacement. A permanent magnet was fixed on the wing while a coil was attached to the base of the generator. The movement of the wing caused the magnetic flux cutting the coil to change, which generated electrical power. The proposed device has dimensions of 12cm × 8cm × 6.5cm. It can start working at a wind speed as low as 2.5m s^{-1} when the generator produced an output power of 470μW. This is sufficient for periodic sensing and wireless transmission. When the wind speed was 5m s^{-1}, the output power reached 1.6mW.

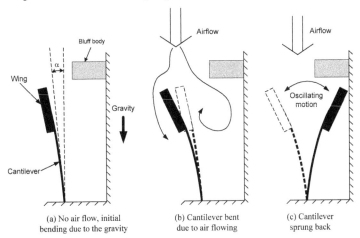

(a) No air flow, initial bending due to the gravity

(b) Cantilever bent due to air flowing

(c) Cantilever sprung back

Fig. 14. Principle of the energy harvester in (Zhu et al., 2010) (transducer is not shown)

4.2 Flutter energy harvesters

The first flapping wind generator was invented by Shawn Frayne and his team in 2004, called Windbelt generator (Windbelt, 2004). The Windbelt generator uses a tensioned membrane undergoing a flutter oscillation to extract energy from the wind as shown in Fig. 15. Magnets are attached to the end of the membrane. They move with the membrane and are coupled with static coils to generate electricity. The company offer Windbelt generators of different sizes. The smallest Windbelt generator has dimensions of 13cm × 3cm × 2.5cm.

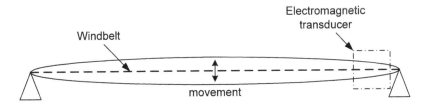

Fig. 15. Windbelt: airflow is perpendicular to this page

The minimum wind speed to make it work is 3m s^{-1}, where an output power less than 100μW was produced. The generator can produce output power of 0.2mW, 2mW and 5mW under the wind of 3.5m s^{-1}, 5.5m s^{-1} and 7.5m s^{-1} respectively (Windbelt, 2004).

Kim et al (2009) reported a small-scale version of the Windbelt generator. The generator had dimensions of 12mm × 12mm × 6mm. The generator was tested under the airflow with the pressure of 50kPa. It produced a voltage output with the frequency of 530Hz and the amplitude of 80mVpp.

Erturk et al (2010) investigated the concept of piezoaeroelasticity for energy harvesting. A mathematical model was established and a prototype device was built to validate the model. The generator had a 0.5m long airfoil vertically placed. Two PZT-5A piezoceramics were attached onto the two ends of the airfoil. Under certain airflow, the airfoil flapped and actuated the piezoceramics to produce electricity. An electrical power output of 10.7mW was delivered to a 100 kΩ load at the linear flutter speed of 9.3m s^{-1}.

Li et al (2009, 2011) reported another type of flapping-leaf which works based on aeroelastic flapping. The device had a PVDF cantilever with its width direction parallel to the air flow. The leaf was placed to make the entire device like an 'L' shape as shown in Fig. 16. Different PVDF cantilevers were compared in the test. It was found that the optimum device generated a peak power of 615μW in the wind of 8m s^{-1}.

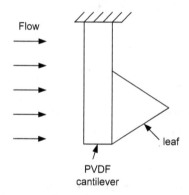

Fig. 16. Flapping-leaf based on aeroelastic flapping

St. Clair et al (2010) reported a micro generator using flow-induced self-excited oscillations. The principle is similar to music-playing harmonicas that create tones via oscillations of reeds when subjected to air blow. Output power between 0.1 and 0.8mW was obtained at wind speeds ranging between 7.5 and 12.5m s^{-1}.

4.3 Energy harvesting with a Helmholtz resonator

4.3.1 Principles

A Helmholtz resonator is a gas-filled chamber with an open neck (as shown in Fig. 17), in which a standard second-order (i.e. spring-mass) fluidic oscillation occurs. The air inside the neck acts as the mass and the air inside the chamber acts as the spring. When air flows past the opening, an oscillation wave occurs. Generally, the cavity has several resonance frequencies, the lowest of which is the Helmholtz resonance. The Helmholtz resonant frequency is given by:

$$f_H = \frac{v}{2\pi}\sqrt{\frac{A}{Vl}} \qquad (6)$$

where v is the speed of sound in a gas, A is the cross sectional area of the neck, l is the length of the neck and V is the static volume of the cavity.

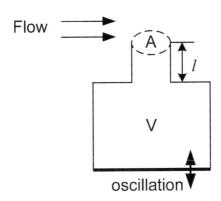

Fig. 17. Helmholtz resonator

4.3.2 Examples

Matova *et al* (2010) reported a device that had a packaged MEMS piezoelectric energy harvester inside a Helmholtz resonator. It was found that packaged energy harvesters had better performance than unpackaged energy harvesters as the package removes the viscous influence of the air inside the Helmholtz cavity and ensure that only the oscillation excites the energy harvester. Experimental results showed that the energy harvester generated a maximum output power of 2µW at 309Hz under the airflow of 13m·s⁻¹. Furthermore, it was found that a major drawback of the Helmholtz resonator is its strong dependence of their resonant frequency on the ambient temperature. This means that this kind of energy harvesters can only be used in the environments with stable temperature or the energy harvester must have a wide operational frequency range.

Kim *et al* (2009) presented a Helmholtz-resonator-based energy harvester with an electromagnetic transducer. The device has a membrane with a magnet attached at the bottom of the cavity. As the membrane oscillates due to the Helmholtz resonance, a static coil is coupled with the moving magnet to generate electricity. Two energy harvesters were fabricated and tested. The first one had dimensions of φ19mm × 5mm and a resonant frequency of 1.4kHz. It generated an open circuit voltage of 4mV$_{pp}$ under the airflow of 0.2kPa (5m·s⁻¹). The second device had dimensions of φ9mm × 3mm and a resonant frequency of 4.1kHz. It generated an open circuit voltage of 15mV$_{pp}$ under the airflow of 1.6kPa.

Liu *et al* (2008) demonstrated the development of an acoustic energy harvester using Helmholtz resonator. It uses a piezoelectric diaphragm to extract energy. The diaphragm consisted of a layer of 0.18mm-thick brass as the substrate and a layer of 0.11mm-thick piezoceramics (APC 850). Experimental results showed an output power of about 30mW was harvested for an incident sound pressure level of 160 dB with a flyback converter. The cavity had dimensions of ϕ12.68mm × 16.4mm.

4.4 Summary

Among these three types of energy harvesters from flow induced vibration, energy harvesters based on VIV and flapping energy harvesters are more suitable for practical application due to their reasonable output power level. Existing energy harvesters with Helmholtz resonators have very low output power and more work needs to be done to make this approach practical. In addition, all piezoelectric flow energy harvesters use PVDF as piezoelectric material due to its flexibility. However, piezoelectric coefficients of PVDF are low compared to those of other piezoelectric materials. Flexible piezoelectric materials with higher piezoelectric coefficients, for example Macro Fiber Composite (MFC), need to be investigated to improve output power of piezoelectric flow energy harvesters.

5. Conclusions

A vibration energy harvester is an energy harvesting device that couples a certain transduction mechanism to ambient vibration and converts mechanical energy to electrical energy. Ambient vibration includes machinery vibration, human movement and flow induced vibration.

For energy harvesting from machinery vibration, the most common solution is to design a linear generator that converts kinetic energy to electrical energy using certain transduction mechanisms, such as electromagnetic, piezoelectric and electrostatic transducers. Electromagnetic energy harvesters have the highest power density among the three transducers. However, performance of electromagnetic vibration energy harvesters reduces a lot in micro scale, which makes it not suitable for MEMS applications. Piezoelectric energy harvesters have the similar power density to the electromagnetic energy harvesters. They have simple structures, which makes them easy to fabricate. Electrostatic energy harvesters have the lowest power density of the three, but they are compatible with MEMS fabrication process and easy to be integrated to chip-level systems.

The linear energy harvester produces a maximum output power when its resonant frequency matches the ambient vibration frequency. Once these two frequencies do not match, the output power drops significantly due to high Q-factor of the generator. Two possible methods to overcome this drawback are tuning the resonant frequency of the generator to match the ambient vibration frequency and widening bandwidth of vibration energy harvesters.

The methods of tuning the resonant frequency include mechanical method and electrical method. The mechanical tuning method requires a certain mechanism to change the mechanical property of the structure of the generator to tune the resonant frequency. Thus, it requires more energy to implement while it normally has a large tuning range.

The electrical tuning method realizes resonant frequency tuning by adjusting electrical loads. This method consumes little energy as it does not involve any change in mechanical properties. In addition, it is much easier to implement than mechanical methods. However, this method normally has a small tuning range.

The suitability of different tuning approaches depends on the application but in general terms, the key factors for evaluating a tuning mechanism are:

- energy consumed by the tuning mechanism should be as small as possible and must not exceed the energy produced by the energy harvester;
- the mechanism should achieve a sufficient operational frequency range;
- the tuning mechanism should achieve a suitable degree of frequency resolution;
- the strategy applied should not increase the damping over the entire operational frequency range.

Energy harvesting from human movement is another important area in vibration energy harvesting. As human movement is random, linear energy harvesters are not suitable for this application. Broadband, non-linear or non-resonant devices are preferred. At the moment, the most common locations on human body for the energy harvesters are feet and upper body due to large displacement or force produced during movement. Up to date, some reported energy harvesters successfully produced useful amount of electrical energy for portable electronic devices. However, consideration needs to be taken to improve design of the energy harvesters so that they will not cause discomfort for human body. Furthermore, another potential solution to energy harvesting from human movement is to print active materials on fabrics, such as jackets and trousers, so that electrical energy can be generated while human body is moving.

Energy harvesters from flow-induced vibration, as an alternative to turbine generators, have drawn more and more attention. Useful amount of energy has been generated by existing devices and the start flow speed has been reduced to as low as 2.5m s^{-1}. However, most reported devices that produce useful energy are too large in volume compared to other vibration energy harvesters. Thus, it is difficult to integrate these devices into wireless sensor nodes or other wireless electronic systems. Future work should focus on miniaturise these energy harvesters while maintain current power level. In addition, researches should be done to further reduce the start flow speed to allow this technology wider applications.

6. References

Allen, J. J. & Smits, A. J. (2001). Energy harvesting eel, In: *Journal of Fluids and Structures*, Vol.15, pp. 629-640, ISSN 0889-9746

Anton, S. R. & Sodano, H. A. (2007). A review of power harvesting using piezoelectric materials (2003-2006), In: *Smart Materials and Structures*, Vol.16, pp.1-21, ISSN 0964-1726

Arakawa, Y.; Suzuki, Y. & Kasagi, N. (2004). Micro seismic power generator using electret polymer film, *Proceedings of PowerMEMS 2004*, 187-190, Kyoto, Japan, November 28-30, 2004

Ayala-Garcia, I. N.; Zhu, D.; Tudor, M. J. & Beeby, S. P. (2009). Autonomous tunable energy harvester, *Proceedings PowerMEMS 2009*, pp. 49-52, Washington DC, USA, December 1-4, 2009

Barrero-Gil, A.; Alonso, G. & Sanz-Andres, A. (2010). Energy harvesting from transverse galloping, In: *Journal of Sound and Vibration*, Vol.329, pp. 2873-2883, ISSN 0022-460X

Beeby, S. P.; Tudor, M. J.; White, N. M. (2006) Energy harvesting vibration sources for microsystems applications, In: *Measurement Science and Technology*, Vol.17, pp. 175-195, ISSN 0957-0233

Beeby, S.; Tudor, M.; Torah, R.; Roberts, S.; O'Donnell, T. & Roy, S. (2007). Experimental comparison of macro and micro scale electromagnetic vibration powered generators, In: *Microsystem Technologies*, Vol.13, No.12-13, pp. 1647-1653, ISSN: 0946-7076

Beeby, S. P.; Torah, R. N.; Tudor, M. J.; Glynne-Jones, P.; O'Donnell, T.; Saha, C. R. & Roy, S. (2007). A micro electromagnetic generator for vibration energy harvesting, In: *Journal of Micromechanics and Microengineering*, Vol.17, pp. 1257-1265, ISSN 0960-1317

Bernitsas, M. M.; Raghavan, K.; Ben-Simon, Y. & Garcia, E. M. H. (2006). VIVACE (Vortex Induced Vibration for Aquatic Clean Energy): A new concept in generation of clean and renewable energy from fluid flow, *Proceedings of OMAE2006 25th International OMAE Conference, Hamburg, Germany, 4-9 June, 2006*

Blevins, R. D. (2001). *Formulas for Natural Frequency and Mode Shape*, Krieger, ISBN 1-57524-184-6, Malabar, Forida, USA

Burrow, S. G. & Clare, L. R. (2007). A Resonant Generator with Non-Linear Compliance for Energy Harvesting in High Vibrational Environments, *IEEE International Electric Machines and Drives Conference*, pp. 715-720, Antalya, Turkey, May 3-5, 2007

Burrow, S. G.; Clare, L. R.; Carrella, A. & Barton, D. (2008). Vibration energy harvesters with non-linear compliance, In: *Active and Passive Smart Structures and Integrated Systems 2008, Proceedings of the SPIE*, Vol.6928, 692807

Cammarano A.; Burrow S. G.; Barton D. A. W.; Carrella A. & Clare L. R. (2010). Tuning a resonant energy harvester using a generalized electrical load, In: In: *Smart Materials and Structures*, Vol.19, 055003(7pp), ISSN 0964-1726

Carroll, D. & Duffy, M. (2005). Demonstration of wearable power generator, *Proceedings of the 11th European Conference on Power Electronics and Applications*, ISBN: 90-75815-09-3, ISBN: 90-75815-09-3, Dresden, Germany, September 11-14, 2005

Charnegie, D. (2007). Frequency tuning concepts for piezoelectric cantilever beams and plates for energy harvesting, *MSc Dissertation*, School of Engineering, University of Pittsburgh, USA

Ching, N. N. H.; Wong, H. Y.; Li, W. J.; Leong, P. H. W. & Wen, Z. (2002). A laser-micromachined vibrational to electrical power transducer for wireless sensing systems, In: *Sensors and Actuators A: Physical*, Vol.97-98, pp. 685-90, ISSN 0924-4247

Despesse, G.; Jager, T.; Chaillout, J.; Leger, J.; Vassilev, A.; Basrour, S. & Chalot, B. (2005). Fabrication and characterisation of high damping electrostatic micro devices for vibration energy scavenging, *Proceedings of Design, Test, Integration and Packaging of MEMS and MOEMS*, pp. 386–390, Montreux, Switzerland, June 1-3, 2005

Dunnmon, J. A.; Stanton, S. C.; Mann, B. P. & Dowell, E. H. (2011). Power extraction from aeroelastic limit cycle oscillations, In: *Journal of Fluids and Structures*, doi:10.1016/j.jfluidstructs.2011.02.003, ISSN 0889-9746

Eichhorn, C.; Tchagsim, R.; Wilhelm, N.; Goldschmidtboeing, F. & Woias, P. (2010). A compact piezoelectric energy harvester with a large resonance frequency tuning range, *Proceedings PowerMEMS 2010*, pp. 207-211, Leuven, Belgium, December 1-3, 2010

Elfrink, R.; Kamel, T. M.; Goedbloed, M.; Matova, S.; Hohlfeld, D.; van Andel,Y. & van Schaijk, R. (2009). Vibration energy harvesting with aluminum nitride-based piezoelectric devices, In: *Journal of Micromechanics and Microengineering*, Vol.19, No.9, 094005, ISSN 0960-1317

Erturk, A.; Hoffmann, J. & Inman, D. J. (2009). A piezomagnetoelastic structure for broadband vibration energy harvesting, In: *Applied Physics Letters*, Vol.94, 254102, ISSN 0003-6951

Erturk, A.; Vieira, W. G. R.; De Marqui, C. Jr. & Inman, D. J. (2010). On the energy harvesting potential of piezoaeroelastic systems, In: *Journal of Applied Physics*, Vol.96, 184103, ISSN 0021-8979

Fang, H. B.; Liu, J. Q.; Xu, Z. Y.; Dong, L.; Wang, L.; Chen, D.; Cai, B. C. & Liu, Y. (2006). Fabrication and performance of MEMS-based piezoelectric power generator for vibration energy harvesting, In: *Microelectronics Journal*, Vol.37, No.11, pp. 1280-1284, ISSN 0026-2692

Feenstra, J.; Granstrom, J.; and Sodano, H. A. (2008). Energy harvesting through a backpack employing a mechanically amplified piezoelectric stack, In: *Mechanical Systems and Signal Processing*, Vol.22, pp. 721–734, ISSN 0888-3270

Ferrari, M.; Ferrari, V.; Guizzetti, M.; Marioli, D. & Taroni, A. (2008). Piezoelectric multifrequency energy converter for power harvesting in autonomous Microsystems, In: *Sensors Actuators A: Physical*, Vol.142, pp. 329–335, ISSN 0924-4247

Ferrari, M.; Ferrari, V.; Guizzetti, M.; Andò, B.; Baglio, S. and Trigona, C. (2009). Improved Energy Harvesting from Wideband Vibrations by Nonlinear Piezoelectric Converters, *Proceedings of Eurosensors XXIII*, Lausanne, Switzerland, September 6-9, 2009

Gieras, J. F.; Oh, J.-H.; Huzmezan, M. & Sane, H. S. (2007). *Electromechanical energy harvesting system*, Patent Publication Number: WO2007070022(A2), WO2007070022(A3)

Glynne-Jones, P.; Tudor, M. J.; Beeby, S. P. & White, N. M. (2004). An electromagnetic, vibration-powered generator for intelligent sensor systems, In: *Sensors and Actuators A: Physical*, Vol.110, pp. 344-349, ISSN 0924-4247

Goldschmidtboeing, F. & Woias, P. (2008). Characterization of different beam shapes for piezoelectric energy harvesting, In: *Journal of Micromachining and Microengineering*, Vol.18, 104013, ISSN 0960-1317

Granstrom, J.; Feenstra, J.; Sodano, H. A. & Farinholt, K. (2007). Energy harvesting from a backpack instrumented with piezoelectric shoulder straps, In: *Smart Materials and Structures*, Vol.16, pp. 1810-1820, ISSN 0964-1726

Hoffmann, D.; Folkmer, B. & Yiannos, M. (2009). Fabrication, characterization and modelling of electrostatic micro-generators, In: Journal of Micromechanics and Microengineering, Vol.19, No.9, 094001, ISSN: 0960-1317

Huesgen, T.; Woias, P. & Kockmann, N. (2008). Design and fabrication of MEMS thermoelectric generators with high temperature efficiency, In: *Sensors and Actuators A: Physical*, Vol.145-146, pp. 423-429, ISSN 0924-4247

Humdinger Wind Energy, LLC, http://www.humdingerwind.com/ (accessable on 13 May 2011).

Jungm, H.-J. & Lee, S.-W. (2011). The experimental validation of a new energy harvesting system based on the wake galloping phenomenon, In: *Smart Materials and Structures*, Vol.20, 055022 (10pp), ISSN 0964-1726

Kim, S.-H.; Ji, C.-H.; Galle, P.; Herrault, F.; Wu, X.; Lee, J.-H.; Choi, C.-A. & Allen, M. G. (2009). An electromagnetic energy scavenger from direct airflow, In: *Journal of Micromechanics and Microengineering*, Vol.19, 094010 (8pp), ISSN 0960-1317

Kok, S.-L.; White, N. M. & Harris, N.R. (2009). Fabrication and characterization of free-standing thick-film piezoelectric cantilevers for energy harvesting, In: Measurement Science and Technology, Vol.20, 124010, ISSN 0957-0233

Koukarenko, E.; Beeby, S. P.; Tudor, M. J.; White, N. M.; O'Donnell, T.; Saha, T.; Kulkani, S. & Roy, S. (2006). Microelectromechanical systems vibration powered electromagnetic generator for wireless sensor applications, In: *Microsystem Technologies*, Vol.12, No.11, pp 1071-1077, ISSN 0946-7076

Kymissis, J.; Kendall, C.; Paradiso, J. & Gershenfeld, N. (1998). Parasitic power harvesting in shoe, Digest of Papers. Second International Symposium on Wearable Computers, PP. 132-139, ISBN 0-8186-9074-7, Pittsburgh, Pennsylvania, USA, October 19-20, 1998

Leland, E. S. & Wright, P. K. (2006). Resonance tuning of piezoelectric vibration energy scavenging generators using compressive axial, In: *Smart Materials and Structures*, Vol.15, pp.1413-1420, ISSN 0964-1726

Li, S. & Lipson, H. (2009). Vertical-stalk flapping-leaf generator for wind energy harvesting, *Proceedings of the ASME 2009 Conference on Smart Materials, Adaptive Structures and Intelligent Systems*, Oxnard, California, USA, September 21-23, 2009

Li, S.; Yuan, J. & Lipson, H. (2011). Ambient wind energy harvesting using cross-flow fluttering, In: *Journal of Applied Physics*, Vol.109, 026104, ISSN 0021-8979

Liao, Y. & Sodano, H. A. (2008). Model of a Single Mode Energy Harvester and Properties for Optimal Power Generation, In: *Smart Materials and Structures*, Vol.17, 065026, ISSN 0964-1726

Liu, F.; Phipps, A.; Horowitz, S.; Ngo, K.; Cattafesta, L.; Nishida, T. & Sheplak, M. (2008). Acoustic energy harvesting using an electromechanical Helmholtz resonator, In: *Journal of Acoustical Society of America*, Vol.123, Vo.4, pp 1983-1990, ISSN: 0001-4966

Lo, H. & Tai, Y. C. (2008). Parylene-based electret power generators, In: *Journal of Micromachining and Microengineering*, Vol.18, 104006, ISSN: 0960-1317

Marzencki, M.; Ammar, Y; & Basrour, S. (2008). Integrated power harvesting system including a MEMS generator and a power management circuit, In: *Sensors and Actuators A: Physical*, Vol.145-146, pp. 363-370, ISSN 0924-4247

Matova, S. P.; Elfrink, R.; Vullers, R. J. M. & van Schaijk, R. (2010). Harvesting energy from airflow with micromachined piezoelectric harvester inside a Helmholtz resonator, *Proceedings of PowerMEMS 2010*, Leuven, Belgium, November 30 - December 3, 2010

MicroBelt tech sheet, http://www.humdingerwind.com/pdf/microBelt_brief.pdf (accessable on 13 May 2011)

Mitcheson, P.; Stark, B.; Miao, P.; Yeatman, E.; Holmes, A. & Green, T. (2003). Analysis and optimisation of MEMS on-chip power supply for self powering of slow moving sensors, *Proceedings of Eurosensors XVII Conference*, pp. 48-51, Guimaraes, Portugal, September 21-24, 2003

Nguyen, D. S. & Halvorsen, E. (2010). Analysis of vibration energy harvesters utilizing a variety of nonlinear springs, *Proceedings of PowerMEMS 2010*, Leuven, Belgium, November 30 - December 3, 2010

Norman, B. C. (2007). Power options for wireless sensor networks, In: *IEEE Aerospace and Electronic Systems Magazine*, Vol.22, No.4, pp. 14-17, ISSN 0885-8985

Peters, C.; Maurath, D.; Schock, W. & Manoli, Y. (2008). Novel electrically tunable mechanical resonator for energy harvesting, *Proceedings PowerMEMS 2008+ µEMS2008*, pp. 253-256, Sendai, Japan, November 9-12, 2008

Ramlan, R. (2009). Effects of non-linear stiffness on performance of an energy harvesting device, *PhD Thesis*, University of Southampton, UK

Renaud, M.; Karakaya, K.; Sterken T.; Fiorini P.; Vanhoof, C. & Puers R. (2008). Fabrication, Fabrication, modelling and characterization of MEMS piezoelectric vibration harvesters, In: *Sensors and Actuators A: Physical*, Vol.145-146, No.1, pp. 380-386, ISSN: 09244247

Peters, C.; Maurath, D.; Schock, W.; Mezger, F. & Manoli, Y. (2008). A closed-loop wide-range tunable mechanical resonator for energy harvesting systems, In: *Journal of Micromechanics and Microengineering*, Vol.19, No.9, 094004(9pp), ISSN 0960-1317

Rome, L. C.; Flynn, L. & Yoo, T. D. (2005) Generating electricity while walking with loads, *Science*, Vol.309, No.5741, (September 2005), pp. 1725-1728, ISSN 0036-8075

Romero, E.; Warrington, R. O. & Neuman, M. R. (2009). Energy scavenging sources for biomedical sensors, In: *Physiological Measurement*, Vol.30, pp. 35-62, ISSN 0967-3334

Roundy, S.; Wright, P. K. & Rabaey, J. (2003). A study of low level vibrations as a power source for wireless sensor nodes, In: *Computer Communications*, Vol.26, pp. 1131-1144, ISSN 0140-3664

Roundy, S. & Wright, P. K. (2004). A piezoelectric vibration based generator for wireless electronics, In: *Smart Materials and Structures*, Vol.13, pp. 1131-1142, ISSN 0964-1726

Roylance, L. & Angell, J. B. (1979). A batch fabricated silicon accelerometer, In: *IEEE Trans. Electron Devices*, Vol.26, pp. 1911–1917, ISSN 0018-9383

Saha, C. R.; O'Donnell, T.; Loder, H.; Beeby, S. P. & Tudor, M. J. (2006). Optimization of an electromagnetic energy harvesting device, In: *IEEE Transactions on Magnetics*, Vol.42, No.10, pp. 3509-3511, ISSN 0018-9464

Saha, C. R.; O'Donnell, T.; Wang, N. & McCloskey P. (2008). Electromagnetic generator for harvesting energy from human motion, In: *Sensors and Actuators A: Physical*, Vol.147, pp. 248–253, ISSN 0924-4247

Sanchez-Sanz, M.; Fernandez, B. & Velazquez, A. (2009). Energy-harvesting microresonator based on the forces generated by the Karman street around a rectangular prism, In: *Journal of Microelectromechanical Systems*, Vol.18, No.2, pp. 449-457, ISSN 1057-7157

Sari, I.; Balkan, T. & Kulah, H. (2007). A wideband electromagnetic micro power generator for wireless Microsystems, *Proceedings of International Solid-State Sensors, Actuators and Microsystems Conference 2007*, pp 275-278, Lyon, France, June 10-14, 2007

Shen, D.; Park, J.; Ajitsaria, J.; Choe, S.; Wikle, H. C. & Kim, D. (2008). The design, fabrication and evaluation of a MEMS PZT cantilever with an integrated Si proof mass for vibration energy harvesting, In: *Journal of Micromechanics and Microengineering*, Vol.18, No.5, 55017, ISSN 0960-1317

Spreemann, D.; Folkmer, B.; Maurath, D. & Manoli, Y. (2006). Tunable transducer for low frequency vibrational energy scavenging, *Proceedings of EurosensorsXX*, Göteborg, Sweden, September 17-20, 2006

St. Clair, D.; Bibo, A.; Sennakesavababu, V. R.; Daqaq, M. F. & Li, G. (2010). A scalable concept for micropower generation using flow-induced self-excited oscillations, In: *Journal of Applied Physics*, Vol.96, 144103, ISSN 0021-8979

Taylor, G. W.; Burns, J. R.; Kammann, S. M.; Powers, W. B. & Welsh, T. R. (2001). The energy harvesting eel: a small subsurface ocean/river power generator, In: *IEEE Journal of Oceanic Engineering*, Vol.26, No.4, pp. 539-547, ISSN 0364-9059

108 Recent Studies in Sustainable Energy Harvesting

Torah, R. N.; Glynne-Jones, P.; Tudor, M. J.; ODonnell, T.; Roy S. & Beeby S. P. (2008). Self-powered autonomous wireless sensor node using vibration energy harvesting, In: Measurement Science and Technology, Vol.19, 125202, ISSN 0957-0233

Mann, B. P. & Owens, B. A. (2010). Investigations of a nonlinear energy harvester with a bistable potential well, In: Journal of Sound and Vibration, Vol.329, pp. 1215-1226, ISSN 0022-460X

von Buren, T. (2006) Body-Worn Inertial Electromagnetic Micro-Generators, PhD Thesis, Swiss Federal Institute of Technology Zurich, Zurich, Switzerland

Wang, D.-A. & Chang, K.-H. (2010). Electromagnetic energy harvesting from flow induced vibration, In: Microelectronics Journal, Vol.41, pp. 356-364, ISSN 0026-2692

Wang, D.-A. & Ko, D.-A. (2010). Piezoelectric energy harvesting from flow-induced vibration, In: Journal of Micromechanics and Microengineering, Vol.20, 025019 (9pp), ISSN 0960-1317

Wang, D.-A. & Liu, N.-Z. (2011). A shear mode piezoelectric energy harvester based on a pressurized water flow, In: Sensors and Actuators A: Physics, Vol.167, No.2, pp. 449-458, ISSN 0924-4247

Wang, D.-A.; Pham, H.-T.; Chao, C.-W. & Chen, J. M. (2011). A piezoelectric energy harvester based on pressure fluctuations in Kármán Vortex Street, Proceedings World Renewable Energy Congress 2011, Linköping, Sweden, 8-11 May 2011

Wang, P.; Dai X.; Zhao X. & Ding G. (2009). A micro electromagnetic vibration energy harvester with sandwiched structure and air channel for high energy conversion efficiency, Proceedings PowerMEMS 2009, pp. 296-299, Washington DC, USA, December 1-4, 2009

Williams, C. B. & Yates, R. B. (1996) Analysis of a micro-electric generator for microsystems, In: Sensors and Actuators A, Vol.52, pp. 8-11, ISSN 0924-4247

Williams, C. B.; Shearwood, C.; Harradine, M. A.; Mellor, P. H.; Birch, T. S. & Yates, R. B. (2001). Development of an electromagnetic micro-generator, In: IEE Proceedings of Circuits Devices Systems, Vol.148, pp. 337-342, ISSN 1350 -2409

Wu, X.; Lin, J.; Kato, S.; Zhang, K.; Ren, T. & Liu, L. (2008). A frequency adjustable vibration energy harvester, Proceedings of PowerMEMS 2008+ µEMS2008, pp. 245-248, Sendai, Japan, November 9-12, 2008

Zhu, D.; Tudor, M. J. & Beeby, S. P. (2010). Strategies for increasing the operating frequency range of vibration energy harvesters: a review, In: Measurement Science and Technology, Vol.21, 022001(29pp), ISSN 0957-0233

Zhu, D.; Roberts, S.; Tudor, J. & Beeby, S. (2010). Design and experimental characterization of a tunable vibration-based electromagnetic micro-generator, In: Sensors and Actuators A: Physical, Vol.158, No.2, pp. 284-293, ISSN 0924-4247

Zhu, D.; Beeby, S.; Tudor, J.; White, N. & Harris, N. (2010). A novel miniature wind generator for wireless sensing applications, Proceedings IEEE Sensors 2010, ISBN 978-1-4244-8170-5, Waikoloa, Hawaii, USA, November 1-4, 2010

Zhu, D.; Almusallam, A.; Beeby, S.; Tudor, J. & Harris, N. (2010). A Bimorph Multi-layer Piezoelectric Vibration Energy Harvester, Proceedings of PowerMEMS 2010, Leuven, Belgium, December 1-3, 2010

Zhu, D. & Beeby, S. (2011). Kinetic energy harvesting, In: Energy Harvesting Systems: Principles, Modeling and Applications, pp. 1-78, Springer, ISBN 978-1-4419-7565-2, New York Dordrecht Heidelberg London

Zhu, D.; Beeby S.; Tudor J. and Harris N. (2011) A credit card sized self powered smart sensor node, In: Sensors and Actuators A: Physical, Vol.169, No.2 pp. 317-325, ISSN 0924-4247

4

Modeling and Simulation of Thermoelectric Energy Harvesting Processes

Piotr Dziurdzia
AGH University of Science and Technology in Cracow
Poland

1. Introduction

Thermoelectric modules are becoming more and more popular nowadays again as their prices are going down and the new potential applications have appeared due to recent developments in microelectronic and wireless technology. Not so long ago Peltier modules were mainly used as thermoelectric coolers TECs, for example in thermal image generator (De Baetselier et al., 1995a), thermoelectrically cooled radiation detectors (Anatychuk, 1995), active heat sinks for cooling of microstructures and microprocessors (Dziurdzia & Kos, 2000), fiber optic laser packages (Redstall & Studd, 1995), special medical and laboratory equipment for temperature regulation (Uemura, 1995), etc. Also in some niche applications, thermoelectric modules working as thermoelectric generators TEGs have been used for some time. Among others, the examples include a miniature nuclear battery for space equipment (Penn, 1974) and remote power stations (McNaughton, 1995).

Fulfilment of the new paradigm Internet of Things (Luo et al., 2009) relating to the idea of ubiquitous and pervasive computing as well as rapid development of wireless sensor networks WSN technologies have attracted recently a great research attention of many R&D teams working in the area of autonomous sources of energy (Paradiso & Starner, 2005), (Joseph, 2005). Apart from light and vibrations, heat energy and thermoelectric conversion are playing an important role in the field of energy harvesting or energy scavenging.

As a rule, thermoelectric generators suffer from relatively low conversion efficiency (not exceeding 12%), so they are practically not applicable to large-scale systems, not to mention power stations. On the other hand they seem to be promising solutions when they are used to harvesting some waste heat coming from industry processes or central heating systems.

In recent years a lot of attention was paid to analyzing Peltier modules and efficiency of thermal energy conversion into electrical one (Beeby & White, 2010), (Priya & Inman, 2009).

Now, many research teams are striving for development of complete autonomous devices powering WSN nodes. Since low power integrated circuits, like microcontrollers, transceivers and sensors, have been commonly available for several years the efforts are focused nowadays especially on ambient energy scavenging and emerging technologies in the field of ultra low voltage conversion, energy storing and efficient power management (Salerno, 2010). There are solutions already reported, operating from extremely low voltages about tens of mV resulting from very small temperature gradients, equaling to single Celsius degrees. In fact, some presented prototypes could be supplied from energy easily available even from human body heat, for example a sensor application (Mateu et al., 2007) and wristwatch (Kotanagi et al., 1999).

Lack of dedicated tools covering complex simulations of thermoelectric devices in both thermal and electrical domains prompted many research teams into developing of original models of Peltier elements facilitating analysis and design of thermoelectric coolers (Lineykin & Ben-Yaakov, 2005), (Dziurdzia & Kos, 1999), (Wey, 2006) as well as thermoegenerators (Chen et al., 2009), (Freunek et al., 2009).

The goal of this text is to show viability of modelling of complex phenomena occurring in thermoelectric devices during energy harvesting as well as coupled simulations both thermal and electrical processes by means of electronic circuits simulators.

Among other benefits such as the low cost, easy to learn notation and built-in procedures for solving differential and nonlinear equations, the electronic circuit SPICE-like simulators have one a very important advantage, namely they are very intuitively understood by electronic engineers community and can be easily used for simulation of other that electrical phenomena. So, the modeling, programming and simulations can be done very fast and in this sway facilitating work of designers. By means of SPICE, provided that a reliable electrothermal model of a Peltier module is available, the energy conversion and distribution flow can be simulated in an autonomous sensor node that is shown in Fig. 1.

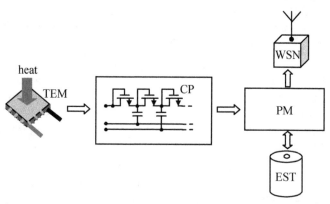

Fig. 1. Thermoelectric energy conversion and distribution flow in an autonomous WSN.

Electric power is produced by a temperature difference between the ambient and the hot surface of a thermoelectric module TEM heated by a waste heat coming from industrial processes, geothermal, isotopic, burned fossil fuels or even human warmth. After that, the generated low voltage is boosted up in a DC/DC converter or a charge pump CP. Next, in power management unit PM the available energy is distributed between autonomous wireless sensor node WSN and the energy storage EST.

A key concern, when designing TEGs for energy harvesters, is not the efficiency but the maximum power transfer to the load. Therefore it is very essential to perform – prior to physical design - series of simulation experiments for different scenarios in order to extract as much as possible electrical power. The presented model is useful in forecasting the operation of TEGs under different conditions relating to temperature as well electrical domains. Even with the best DC/DC converter boosting up the voltage to supplying an electronic circuitry one has to remember that the thermoelectric energy harvesting is a low efficiency method and there is not much power available. Therefore a lot of effort should be invested in simulation and design stage of energy harvesters based on Peltier modules.

In the next following paragraphs basics of thermoelectric modules based on Peltier devices are shown, with the phenomena that rule their operation and are crucial for comprehensive understanding of heat to electric energy conversion. After that an analytical description of the heat flux and power generation in TEMs is presented, followed by electrothermal modelling in electronic circuit simulator. At the end a set of simulations scenarios for thermogenerator based on a commercially available thermoelectric module is shown. The results of simulations experiments are very useful in predicting maximum ratings of the TEGs during operation under different ambient conditions and electrical loads.

2. Basics of thermoelectric generators

Thermoelectric generator TEG is a solid-state device based on a Peltier module, capable of converting heat into electrical energy. In the opposite mode of work when it is supplied with DC current it is able to pump heat, which in consequence leads to cooling one of its sides whereas heating of the other The Peltier module consists of N pairs of thermocouples connected electrically in series and thermally in parallel. They are sandwiched between two ceramic plates which are well conducting heat but on the other hand representing high electrical resistance (Fig. 2).

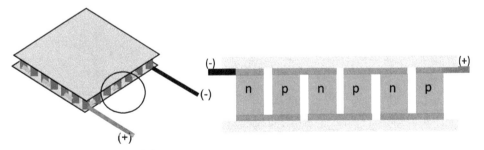

Fig. 2. Thermoelectric module.

Thermoelectric material is characterized by the figure of merit Z which is a measure of its suitability for thermoelectric applications (1). Good materials should have high Seebeck coefficient a, low electrical resistivity ρ and low thermal conductivity λ.

$$Z = \frac{\alpha^2}{\rho\lambda} \tag{1}$$

The most commonly used thermocouples in modules are made of heavily doped bismuth telluride Bi_2Te_3. They are connected by thin copper strips in meander shape and covered by two alumina Al_2O_3 plates.

The overall operation of a TEG is governed by five phenomena, i.e.: Seebeck, Peltier, Thomson, Joule and thermal conduction in the materials. Some of them foster thermoelectric conversion but a few of them limit the TEG performance.

2.1 Seebeck effect

Seebeck Effect describes the induction of a voltage V_S in a circuit consisting of two different conducting materials, whose connections are at different temperatures. In case of a Peltier module the Seebeck voltage can be expressed as in (2), where T_h-T_c is the temperature gradient across the junctions located at the opposite sides of the module.

$$V_S = \alpha\left(T_h - T_c\right) \tag{2}$$

2.2 Peltier effect

Peltier phenomenon describes the processes occurring at the junction of two different conducting materials in the presence of a flowing electrical current. Depending on the direction of current flow the junction absorbs or dissipates heat to the surroundings. The amount of absorbed or dissipated heat is proportional to the electrical current and the absolute temperature T. The heat power associated with the Peltier phenomenon can be calculated as in (3),

$$Q_P = \pi I = \alpha T I \tag{3}$$

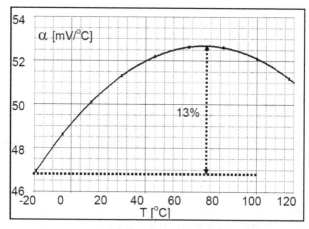

Fig. 3. Seebeck coefficient against temperature.

where I is the electrical current flowing in the thermoelectric module, π is the Peltier coefficient that can be expressed by means of Seebeck coefficient a. For bismuth telluride, Seebeck coefficient is not constant but slightly temperature dependent. In Fig. 3, function of the Seebeck coefficient against temperature, for a commercially available thermoelectric module is shown.

Peltier effect is the basis of the thermoelectric coolers, while the Seebeck effect is used in electrical power generators.

2.3 Thomson effect

Thomson phenomenon takes place in presence of an electrical current flowing not through a junction of two materials as in Peltier effect but in a homogeneous electrical conductor placed between objects at two different temperatures. Depending on the direction of current flow, a heat is absorbed or dissipated from the conductor volume. For instance, if the electrons are the current carriers and move towards higher temperatures, in order to maintain thermal equilibrium they must take an energy as heat from the outside. The reverse situation occurs in the opposite direction of the current flow. Quantitative model of this effect is described by (4) (Lovell et al., 1981),

$$Q_t = -\mu_T \cdot I \cdot \frac{dT}{dx} \tag{4}$$

where μ_T is the Thomson coefficient.

The influence of Thomson effect on performance of thermoelectric devices is very weak, however it exists and cannot be neglected for very high temperature gradients.

2.4 Joule heat phenomenon

Joule heat generation is the most commonly known phenomena associated with a current flowing in electrical circuits. Opposite to the previously described phenomena, Joule effect is not reversible and it manifests in a heat dissipated by material with non-zero resistance in the presence of electrical current (5).

$$Q_j = I^2 \cdot R \tag{5}$$

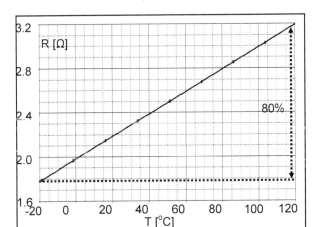

Fig. 4. Internal resistance of a thermoelectric module against temperature.

In Fig. 4, a temperature function of the internal resistance of a thermoelectric module is shown.

2.5 Heat conduction

Heat flow and conduction between two sides of a thermoelectric module is described in details in the next paragraph. An important difficulty in describing this phenomenon in the case of Peltier modules is a significant temperature difference across the active material of Bi_2Te_3 and more over the strong temperature dependence of the thermal conductivity K, as shown in Fig. 5.

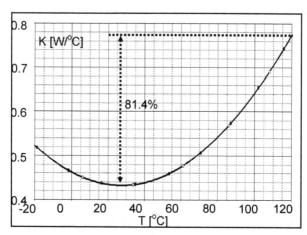

Fig. 5. Thermal conductivity of a thermoelectric device against temperature.

2.6 Power generation

When a thermoelectric couple or a meander of serially connected pairs is placed between two objects at two different temperatures T_c and T_h - e.g. a heat sink and a heat source - it can produce Seebeck voltage V_S (Fig. 6). In this case only Seebeck effect and heat conduction phenomenon occur.

If the electromotive force V_S is closed by a resistive load R_L then an electrical power P is generated (6) and the thermoelectric module is utilizing all the described phenomena.

$$P = I^2 R_L = \left(\frac{V_S}{R_L + R_I} \right)^2 R_L = \left(\frac{\alpha (T_h - T_c)}{R_L + R_I} \right)^2 R_L \qquad (6)$$

Where, R_I is the internal resistance of the thermoelectric couples made of bismuth telluride.

2.7 Benefits of thermoelectric generators

Thermoelectric modules manifest some advantages when the other harvesting methods and sources of energy coming from the environment are considered. First of all, thermoelectric generation is some kind of solid state power conversion. Therefore the Peltier devices do not have any moving parts, so they are reliable, silent and they are characterized by very long MTF (mean time to failure). Moreover they are not chemically hazardous. Next, opposite to photovoltaic panels they can operate in conditions where there light is not sufficient or not available at all. Finally, temperature gradients have tendency to change rather more slowly than the amplitudes of vibrations which often are occurring as single bursts. Therefore, thermoelectric generators can provide energy in a continuous way.

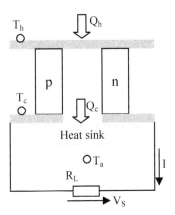

Fig. 6. Power generation by a single thermocouple exposed to a temperature gradient.

3. Analytical analysis of thermoelectric devices

During considerations on modeling of thermoelectrical energy processes generation one has to take into account electrothermal interactions between a few phenomena that form a feedback loop as depicted in Fig. 7.

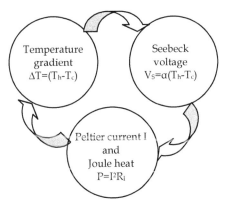

Fig. 7. Electrothermal interactions in thermoelectric modules working as TEGs.

A temperature gradient ΔT resulting from different ambient conditions between two sides of Peltier module causes that a Seebeck voltage V_S appears. If the circuit is closed by a certain resistive load R_L, the voltage V_S forces a Peltier current I flow, and in consequence there appears a Joule heat resulting from dissipated power in the internal resistance R_I of the Peltier module. Joule heat introduces some temperature disturbance to the existing temperature gradient, and thus influences on the Seebeck voltage. Then the whole cycle starts again.

In order to derive quantitative description of the TEG operation a layered model will be analysed which is shown in Fig. 8. The passive elements of the TEG will be described by means of the general equation of heat conduction (7), while the active parts will be modeled according to the constant parameters theory (Buist, 1995).

$$\lambda \nabla T(x,y,z,t) + w(x,y,z,t) = C_9 \frac{\partial T(x,y,z,t)}{\partial t} \tag{7}$$

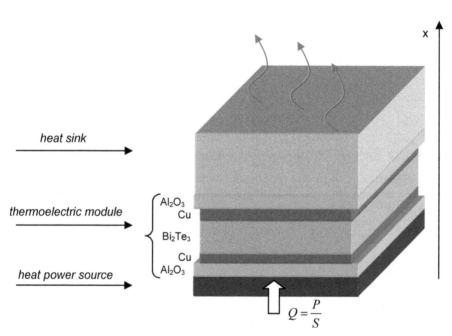

Fig. 8. Layered model of a thermoelectric generator subjected to analysis.

Where, w is generated heat power density distribution, C_9 is the specific heat capacity coefficient.

3.1 Heat conduction in passive layers of a thermoelectric module

With a good approximation it can be assumed one-directional heat flow due to much larger planar dimensions of the thermoelectric generator than the lateral ones. It means that the surfaces that are parallel to the direction of heat flow can be treated as adiabatic ones (8) (De Baetselier et al., 1995b).

$$\frac{\partial T(x,y,z)}{\partial y} = 0, \qquad \frac{\partial T(x,y,z)}{\partial z} = 0 \qquad (8)$$

Differential equation for a heat flow in a steady state, without internal heat sources, can be expressed by (9). In (10) and (11), boundary conditions for interfaces between heat source and Al_2O_3 as well as Al_2O_3 and copper strips are presented.

$$\frac{d^2T}{dx^2} = 0 \qquad (9)$$

$$\left.\frac{dT}{dx}\right|_{x=0} = \frac{P}{S_{Al_2O_3} \cdot \lambda_{Al_2O_3}} \qquad (10)$$

$$\lambda_{Al_2O_3} \left.\frac{dT_{Al_2O_3}(x)}{dx}\right|_{x=l_{Al_2O_3}} = \lambda_{Cu} \left.\frac{dT_{Cu}(x)}{dx}\right|_{x=0} \qquad (11)$$

Finally, for the galvanic connection between copper layer and the cold side of the bismuth telluride, the temperature is equal to T_h – temperature of the hot side of the active part of the TEG (12).

$$\left.T_{Cu}(x)\right|_{x=l_{Cu}} = T_h \qquad (12)$$

For the opposite side of the thermoelectric modules we can derive similar equations, except that the Al_2O_3 layer at the cold surface is adjacent to a heat sink (13).

$$\lambda_{Al_2O_3} \left.\frac{dT_{Al_2O_3}(x)}{dx}\right|_{x=l_{Al_2O_3}} = \lambda_{hs} \left.\frac{dT_{hs}(x)}{dx}\right|_{x=0} \qquad (13)$$

The other side of the heat sink is exposed to an ambient temperature Ta. The heat is transferred to the surrounding environment by radiation and convection which are described by the average heat transfer coefficient h (14) (Kos, 1994).

$$\left.\frac{dT}{dx}\right|_{x=hs} = -\frac{\bar{h}}{\lambda_{hs}}(T_{hs}-T_a) \qquad (14)$$

3.2 Heat flow and power generation in active part of a thermoelectric module

According to the thermoelectric theory based on constant parameters the active part of a thermoelectric generator can be described by a set of three equations. Two of them are relating to the thermal domain and represent heat powers Q_c at the cold side (15), and Q_h at the hot one (16), while the last one comes from the electrical domain and represents an electrical circuit consisting of an electromotive force V_S causing a Peltier current I flow (17).

$$Q_c = \alpha_{Bi_2Te_3}(T) \cdot T_c \cdot I - \frac{I^2 \cdot R_{Bi_2Te_3}(T)}{2} - K_{Bi_2Te_3}(T) \cdot (T_h - T_c) = Q_{c1} - Q_{c2} - Q_{c3} \qquad (15)$$

$$Q_h = \alpha_{Bi_2Te_3}(T) \cdot T_h \cdot I + \frac{I^2 \cdot R_{Bi_2Te_3}(T)}{2} - K_{Bi_2Te_3}(T) \cdot (T_h - T_c) = Q_{h1} + Q_{h2} - Q_{h3} \qquad (16)$$

$$V_S = \left[\alpha_{Bi_2Te_3}(T) \cdot T_h - \alpha_{Bi_2Te_3}(T) \cdot T_c \right] \qquad (17)$$

Neglecting the Thomson effect, the thermoelectric device can be shown as two heat power generators (Fig. 9) consisting of components responsible for Peltier effect Q_{c1} and Q_{h1}, Joule heat Q_{c2} and Q_{h2}, heat conduction Q_{c3} and Q_{h3}. In case of the Joule heat it is assumed that one half of it dissipates at the cold side and the other half flows to the other side of the thermoelectric generator.

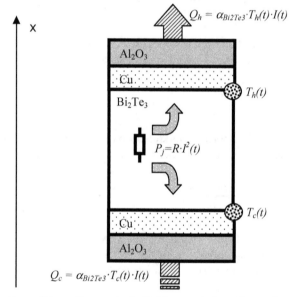

Fig. 9. Cross section of the active part of a Peltier module with two heat power generators.

4. Electrothermal model of thermoelectric generator based on Peltier modules

Complexity of the electrothermal behaviour of thermoelectric devices - that is described by nonlinear differential equations - can be represented and solved by means of finite element modeling (FEM). However, such a sophisticated tool is impractical from the electronic engineers' point of view who need intuitively easy to understand and user friendly simulators.

An electrothermal model of a TEG based on Peltier module makes possible for engineers to carry out investigations - not necessarily going into physical details - on free power generators converting some available ambient energy. To properly model electro-thermal phenomena in SPICE-like simulators the well known analogy has to be used, where: temperature T corresponds to electrical voltage V, heat flux Q to electrical current I, and thermal resistance R_{th} to electrical resistance R.

Due to the fact that the SPICE-like simulators are dedicated to simulations of only electric phenomena, the whole synthesised equivalent circuit representing all the physical aspects of a TEG operation has to be separated into two different circuits: one representing thermal phenomena and the other for electrical ones (Fig. 10). These two circuits are in mutual interaction by exchanging information about generated heat flux Q and changing temperature T. The real electrical circuit represents an electromotive force the V_S that causes a Peltier current flow and in consequence Joule heat generation. The heat flux Q is converted into thermal domain by means of an auxiliary thermal circuit and influences on the operation of the thermal model of the TEG that is represented by a real thermal circuit. Next, it introduces some changes in temperature distribution which in turn, by means of an auxiliary electrical circuit, modifies the Seebeck voltage. By means of the electrothermal analogy, only equivalent electrical elements are needed to be used for modelling of heat conduction as well as power heat generation.

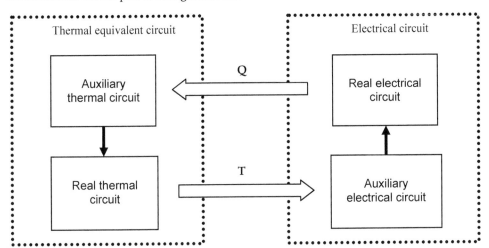

Fig. 10. Block diagram illustrating interaction between the electrical circuit and the thermal equivalent circuit.

4.1 Synthesis of the equivalent electrothermal model of a TEG

By solving the set of equations (9)-(12) describing heat conduction in passive layers of the TEG, the formulas for the hot and cold sides respectively are obtained in (18) and (19) (Janke, 1992). They can be treated as Kirchhoff's law in thermal domain.

$$T_{heat\_source} = T_h + P\left(\frac{l_{Al_2O_3}}{S_{Al_2O_3} \cdot \lambda_{Al_2O_3}} + \frac{l_{Cu}}{S_{Cu} \cdot \lambda_{Cu}}\right) = T_h + P\left(R_{th\_Al_2O_3} + R_{th\_Cu}\right) \qquad (18)$$

$$T_c = T_a + Q_c \left(R_{th\_hs} + R_{th\_Al_2O_3} + R_{th\_Cu} \right) \tag{19}$$

R_{th_hs} is the thermal resistance represented by a heat sink and it takes into account heat conduction, convection as well as radiation (20).

$$R_{th\_hs} = \frac{l_{hs}}{S_{hs} \cdot \lambda_{hs}} + \frac{1}{h \cdot S_{hs}} \tag{20}$$

Appropriate model should also describe the behaviour of the TEG in transient states. For this purpose a concept of thermal capacity C_{th} is introduced. Interpretation of the thermal capacity is obtained by comparing the temperature impulse response characteristics for a single layer with the analytical solution of this problem. As a result, the thermal capacity for a single layer can be expressed as in (21) (Janke, 1992).

$$C_{th} = \frac{\tau}{R_{th}} = \frac{4 \cdot C_g \cdot l \cdot S}{\pi^2} \tag{21}$$

From equations (18)-(21) one can easily derive equivalent passive elements of the TEG that form an equivalent circuit. After connection the active heat power sources and Seebeck voltage described in (15)-(17) a complete model is obtained (Fig. 11).

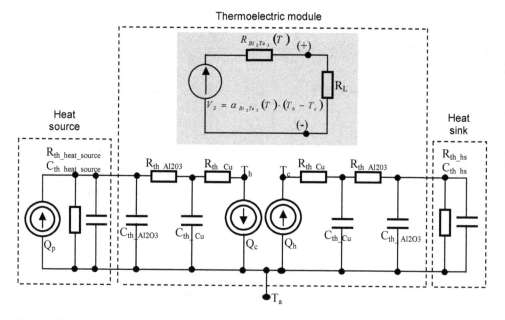

Fig. 11. Complete equivalent electrothermal model of a thermoelectric generator.

The final model of the TEG was converted to a netlist and implemented into electronic circuits simulator. To get products of some quantities that constitute Q_c, Q_h and V_s, auxiliary voltage dependent current sources were used.

```
Gamc 0 11 POLY(1) 100 0 11.03m 28.6u -188n
Rtym_amc 11 0 1
Gqc1 100 500 POLY(3) 11 0 110 0 300 0 0 0 0 0 0 0 0 0 0 0 0 0 0 1
Grmc 0 22 POLY(1) 100 0 352m 1.7m
Rtym_rmc 22 0 1
Gqc2 500 100 POLY(2) 22 0 300 0 0 0 0 0 0 0 0 0 0.5
Gkmc 0 33 POLY(1) 100 0 174m 290u 7.25u
Rtym_kmc 33 0 1
Gqc3 500 100 POLY(3) 33 0 200 0 100 0 0 0 0 0 0 1 -1
```

Fig. 12. A part of a SPICE notation describing the cold heat power source.

In the SPICE notation, the final model of the TEG is seen as a subcircuit consisting of four terminals, as shown in Fig. 13.

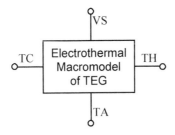

Fig. 13. Macromodel of TEG.

Nodes TC and TH represent temperatures in ºC at the cold and hot sides respectively, while TA refers to an ambient temperature. These three input terminals come from thermal equivalent circuit and during simulations in SPICE they are seen as voltages. They can be connected to the voltage sources (any function) and then the conditions would be similar to those as the TEG was placed between two objects of infinite thermal capacity. They can also be connected to the passive thermal circuits or the heat sources, of any time functions, that are represented in equivalent circuits by current sources. The terminal VS comes from electrical part of the model. It is the output node where Seebeck voltage appears. It can be connected to any kind of a resistive load to simulate real conditions of energy harvesting processes.

4.2 Improved electrothermal model

Assumption about constant parameters a, R, K seems to be a simplified solution in view of the temperature range the TEG is going to work in, and because of the actual temperature dependency of the parameters, because they show strong nonlinear temperature dependency as it was presented in Fig. 3 – Fig. 5. In typical working conditions, common in the industrial environment, for temperatures ranging from -20ºC to 120ºC, the parameters can change their values from 13% (for a) beyond 80% (for K)!

To increase the accuracy of the model some modifications were made. Among others, the dependencies of the coefficients: α, R, K, on temperature were introduced. Their changes with temperature were taken into consideration and implemented into the model by approximation of polynomial second order functions. It was assumed that the temperature argument T appearing in equations (15)-(17) should match the average temperature T_{av} between the two sides of the thermoelectric module (Seifert et al., 2001).

4.3 Experimental characterization of thermoelectric modules

Unfortunately, temperature characteristics of the coefficients of thermoelectric modules are not disclosed to the users by manufacturers, as a general rule. Due to this fact, before they can be taken advantage of, in constructing of an electrothermal model, first they need to be calculated during experimental characterization processes.

Methodology for extracting thermoelectric module parameters can be found in the work (Mitrani et al., 2005). Slightly different approach was demonstrated by (Dalola et al., 2008).

To determine temperature dependences of Seebeck coefficient α_{Bi2Te3} and electrical resistance of thermoelectric couples R_{Bi2Te3} the TEM under test should be placed between two objects of controlled in a wide range temperatures T_c and T_h. From measured output voltage V_O and electrical current I the two parameters can be derived as in (22) and (23).

$$\alpha_{Bi_2Te_3} = \frac{V_O}{T_h - T_c}\bigg|_{I=0} \tag{22}$$

$$R_{Bi_2Te_3} = \frac{V_O|_{I=0} - V_O}{I} \tag{23}$$

Experimental measurement of the thermal conductivity K_{Bi2Te3} brings about much more problems because to its determination a heat flux Q should be known. Therefore, some modification in measurement setup which is shown in Fig. 14 is necessary.

Temperature of the cold side of the thermoelectric module under test is controlled in a closed feedback loop by an auxiliary thermoelectric cooler. The other side of the module is fixed to an auxiliary heat conducting block made of aluminum or copper, with adiabatic surfaces. The other end of the block is fixed to an auxiliary TEM with controlled temperature T_x, which is higher than T_c. Provided that the thermal conductivity coefficient of the auxiliary conducting block is known, the heat flux is easily obtained as in (24).

$$Q_h = Q_x = \lambda_{Al}(T_h - T_c) \tag{24}$$

Taking into account the equation (14), thermal conductivity for bismuth telluride active part can be calculated as in (25).

$$K_{Bi_2Te_3} = \frac{\alpha_{Bi_2Te_3} T_h I + 0.5 I^2 R_{Bi_2Te_3} - \alpha_{Al}(T_h - T_x)}{T_h - T_c} \tag{25}$$

The obtained temperature characteristics of the main three parameters are approximated by polynomial second order functions, which for the examined Peltier device were shown in

Fig. 3 – Fig. 5. The coefficients staying before variables in the approximated functions are next introduced to the core netlist notation of the SPICE-like simulators.

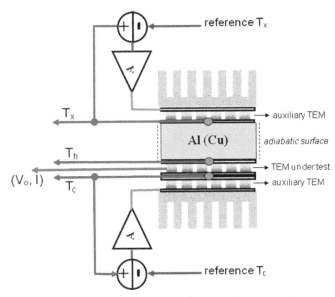

Fig. 14. Measurement setup for characterization of thermoelectric modules.

5. Simulations of energy harvesting processes in thermoelectric generators

The crucial part in the whole design process of TEGs is the modeling and electrothermal simulations which can provide good estimations of electrical energy that can be obtained from heat conversion. Properly designed generator requires many simulations during design stage against different conditions, so as to obtain as much as possible electrical energy from a relatively low efficiency system. In some cases it can replace prototyping of complex designs. The implemented macromodel of TEG into SPICE-like simulators makes possible to perform different simulation scenarios. In the next following examples some results of simulations experiments that were performed by means of the electrothermal model of TEG and electronic circuits simulator SPICE are presented. They give quantitative information about behaviour of the real TEG under different thermal and electrical conditions often occurring in practice. All the presented simulations were performed for a 127 pairs commercially available thermoelectric module for which the approximated polynomial functions look like in (26)-(28).

$$\alpha_{Bi_2Te_3}(T) = -7 \cdot 10^{-7} T^2 + 1 \cdot 10^{-4} T + 49.1 \cdot 10^{-3} \tag{26}$$

$$R_{Bi_2Te_3}(T) = 10.1 \cdot 10^{-3} T + 1.98 \tag{27}$$

$$K_{Bi_2Te_3}(T) = 4 \cdot 10^{-5} T^2 - 2.2 \cdot 10^{-3} T + 0.4626 \tag{28}$$

5.1 Examples of simulations

Output voltage V_S from the TEG for $R_L=\infty$ is shown in the Fig. 15. The simulations were performed for different temperature gradients by setting T_c as parameter and T_h as an argument. A scheme of the equivalent circuit for the presented simulations is shown next to the characteristics.

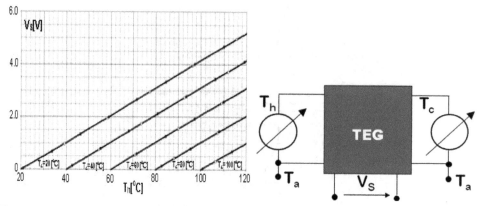

Fig. 15. Output voltage V_S versus temperature of the hot side and constant T_C.

The graphs allow estimating maximum ratings of voltages when no resistive load is present. They can be very useful in the course of designing DC-DC boost regulators pushing up the voltage to the level required by an electronic circuitry, for example sensor nodes.

In the Fig. 16 an output power P_L in relation to a resistive load R_L and different temperature gradients ΔT is depicted. From the simulated functions one can find the points of the best matching between R_L and the inner resistance of the Peltier module. It is worth mentioning that the maximum power transfer point (MPTP) is not constant but moves in the direction of higher R_L as the temperature gradient ΔT is increasing.

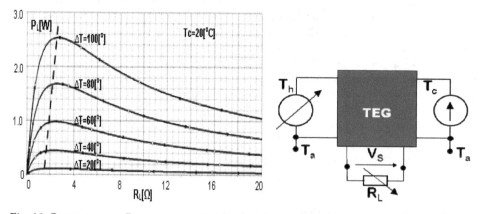

Fig. 16. Output power P_L versus resistive load and constant temperature gradient ΔT.

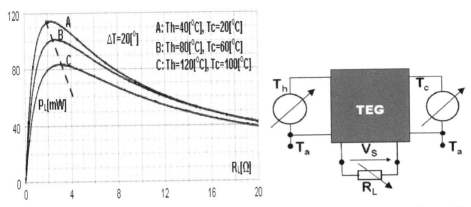

Fig. 17. Output power P_L versus resistive load R_L and the same temperature gradients ΔT, located in low, medium and high temperatures.

Next, the Fig. 17 shows the output power P_L in the function of R_L for the same $\Delta T = 20^\circ$, in low, medium and high temperatures. From this picture it is evident to designers that the maximum output power P_L depends not only on R_L and ΔT, but also on the temperature range the ΔT gradient is located in.

Functions of coefficient of performance COP, defined as in (29), are shown in Fig. 18 and Fig 19. They both present conversion efficiency in relation to a resistive load R_L and a heat power Q_h. It can be observed that COP does not exceed 12%.

$$COP = \frac{P_L}{Q_h} \cdot 100\% \tag{29}$$

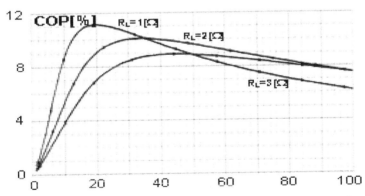

Fig. 18. Coefficient of performance COP versus heat power of the hot side and constant R_L.

Results presented in Fig. 20 illustrate importance of proper selection of the heat sink that is attached to the cold side of the module on the overall performance of TEG. From the point of view of the output voltage, it seems that the thermal resistance R_{th_hs} of the heat sink should be chosen as low as possible, so that the high heat power Q_h would resulted in a large enough temperature gradient across the Peltier module. Otherwise, too high R_{th_hs} might squander the whole effort that was made to improve the overall system efficiency. On the other hand it can be proved that similarly to the idea of electrical matching, the maximum power transfer can be provided if the thermal resistance of the heat sink is matched with thermal resistance of the thermoelectric module.

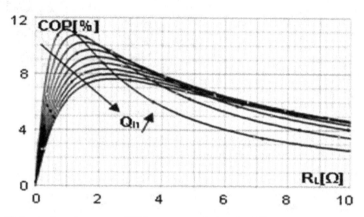

Fig. 19. Coefficient of performance COP versus resistive load R_L.

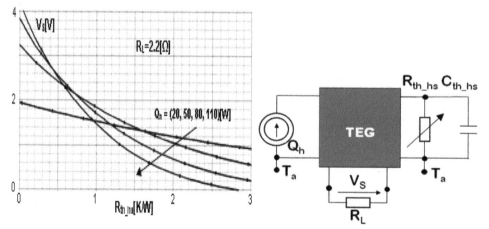

Fig. 20. Output voltage V_S versus thermal resistance of a heat sink.

6. Conclusion

Thermoelectric generators fulfill a very important role in some niche applications, for example they provide energy for satellites and space equipment. Smaller TEGs can supply electrical energy to remote distant low power consumption objects and when there is no electricity nearby. Thermoelectric modules applied as electrical energy generators can be treated as potential sources of green energy, although as a matter of fact they suffer from relatively low efficiency so far. Hopefully, with new materials technology development TEGs will have better energy conversion coefficients of performance in the future. However, it is obvious that for energy harvesting applications Peltier devices can have a very high potential impact.

Presented electrothermal model took into account both thermal and electrical phenomena taking place in Peltier modules. The model is easy to adapt in any electronic circuits simulator, especially in SPICE which has well established position as convenient and reliable software among electronic engineers. The improved version of the model took also the temperature dependent coefficients, describing operation of the TEGs, into consideration, thus leading to its increased accuracy.

Obtained experimental simulations give a number of interesting insights into interactions between TEG, surrounding environment and the electrical load. In this way, they make possible to perform qualitative analysis of energy harvesting processes. They proved the usefulness of the model for designers of the real harvesters made of Peltier modules, thanks to the fact that the whole design process is then more cost effective and efficient.

Future research work will be focused on more accurate modeling of temperature functions of the crucial parameters. In fact the average temperature T_{av} that is used in the improved model for determining actual values of the coefficients should be calculated by integration of the nonlinear temperature profiles of coefficients along the pellets. This issue should be taken into account especially for high temperature gradients.

7. Acknowledgement

The work was supported by the National Centre for Research and Development (NCBiR) project grant No. R02 0073 06/2009.

8. References

Anatychuk L. I. (1995). Thermoelectrically Cooled Radiation Detectors, *CRC Handbook of Thermoelectrics*, CRC Press 1995, pp. 633-640.

Beeby S., White N. (2010). Energy Harvesting for Autonomous Systems, *Artech House*, 2010, ISBN-13: 978-1-59693-718-5.

Buist R. J. (1995). Calculation of Peltier device performance, *CRC Handbook of Thermoelectrics*, CRC Press 1995, pp. 143-155.

Chen M., Rosendahl L. A., Condra T. J., Pedersen J. K. (2009). Numerical Modeling of Thermoelectric Generators With Varing Material Properties in a Circuit Simulator, *IEEE Transactions on Energy Conversion*, Vol. 24, No. 1, March 2009, pp. 112-124.

Dalola S., Ferrari M., Ferrari V., Guizzetti D. M., Taroni A. (2008). Characterization of Thermoelectric Modules for Powering Autonomous Sensors, *IEEE Transactions on Instrumentation and Measurement*, Vol. 58, No. 1, January 2009, pp. 99-107.

De Baetselier E., De Mey G., Kos A. (1995a). Thermal Image Generator as a Vision Prosthesis for the Blind, *MST Poland News*, No. 3(7), October 1997, pp. 3-5.

De Baetselier E., Goedertier W., De Mey G. (1995b). Thermoregulation of IC's with high power dissipation, *Proceedings of the 10th European hybrid Microelectronics Conference*, Copenhagen, Denmark, May 1995, ISHM, pp. 425-439.

Dziurdzia P., Kos A. (1999). Electrothermal Macromodel of Active Heat Sink for Cooling Process Simulation, *Proc. of the 5-th International Workshop on Thermal Investigations of Ics and Microstructures*, Roma, Italy, October 4-6, 1999, pp.76-81.

Dziurdzia P., Kos A. (2000). High Efficiency Active Cooling System, *Proceedings. of the XVIth Annual IEEE Semiconductor Thermal Measurement and Management Symposium SEMITHERM,* San Jose, USA, 21-23 March 2000, pp. 19-26.

Freunek M., Muller M., Ungan T., Walker W., Reindl L. M. (2009). New Physical Model for Thermoelectric Generators, *Journal of Electronic Materials,* Vol. 38, No. 7, 2009, pp. 1214-1220.

Janke W. (1992). Zjawiska termiczne w elementach i układach półprzewodnikowych, *Układy i systemy elektroniczne, Wydawnictwa Naukowo - Techniczne,* Warszawa 1992.

Joseph A. D. (2005). Energy Harvesting Projects, *Published by the IEEE CS and IEEE ComSoc,* 1536-1268/05.

Kos A. (1995). Modelowanie hybrydowych układów mocy i optymalizacja ich konstrukcji ze względu na rozkład temperatury, *Rozprawy Monografie, Wydawnictwa AGH,* Kraków 1994.

Kotanagi S. et al. (1999). Watch Provided with Thermoelectric Generator Unit, Patent No. WO/1999/019775.

Lineykin S., Ben-Yaakov S. (2005). Analysis of thermoelectric coolers by a spice-compatible equivalent-circuit model, *Power Electronics Letter IEEE,* Volume 3, Issue 2, 2005, pp. 63-66.

Lovell M. C., Avery A. J., Vernon M. W. (19821). Physical properties of materials, *Van Nostrand Reinhold Company, University Press,* Cambridge 1981.

Luo J., Chen Y., Tang K., Luo J. (2009). Remote monitoring information system and its applications based on the Internet of Things, *BioMedical Information Engineering, 2009. FBIE 2009. International Conference on Future ,* 13-14 Dec. 2009, pp.482-485.

Mateu L., Codrea C., Lucas N., Pollak M., Spies P. (2007). Human Body Energy Harvesting Thermogenerator for Sensing Applications, *Proc. of the International Conference on Sensor Technologies and Applications SensorComm 2007,* October, Valencia, Spain, pp. 366-372.

McNaughton A. G. (1995). Commercially Available Generators, *CRC Handbook of Thermoelectrics,* CRC Press 1995, pp. 659-469.

Mitrani D., Tome J. A., Salazar J., Turo A., Garcia M. J., Chavez A. (2005). Methodology for Extracting Thermoelectric Module Parameters, *IEEE Transactions on Instrumentation and Measurement,* Vol. 54, No. 4, August 2005, pp. 1548-1552.

Paradiso J. A., Starner T. (2005). Energy Scavenging for Mobile and Wireless Electronics, *Pervasive Computing, IEEE,* Jan.-March 2005, pp. 18-27.

Penn A. (1974). Small electrical power sources, *Phys. Technol,* 5, 114,1974.

Priya S., Inman D. J. (2009). Energy Harvesting Technologies, *Springer,* 2009, ISBN 978-0-387-76463-4.

Redstall R. M., Studd R. (1995). Reliability of Peltier Coolers in Fiber-Optic Laser Packages, *CRC Handbook of Thermoelectrics,* CRC Press 1995, pp. 641-645.

Salerno D. (2010). Ultralow Voltage Energy Harvester Uses Thermoeletric Generator for Battery-free Wireless Sensors, *LT Journal,* 2010, pp. 1-11.

Seifert W., Ueltzen M., Strumpel C., Heiliger W., Muller E. (2001). One-Dimensional Modeling of a Peltier Element, *Proc. of the 20th International Conference on Thermoelectrics,* 2001.

Uemura K. (1995). Laboratory Equipment, *CRC Handbook of Thermoelectrics,* CRC Press 1995, pp. 647-655.

Wey T. (2006). On the Behavioral Modeling of a Thermoelectric Cooler and Mechanical Assembly, *IEEE North-East Workshop on Circuit and Systems,* 2006, pp. 277-280.

Part 2

Future:
Sustainable Energy Harvesting Techologies

5

Energy Harvesting Technologies: Thick-Film Piezoelectric Microgenerator

Swee Leong Kok
Faculty of Electronic and Computer Engineering,
Universiti Teknikal Malaysia Melaka,
Malaysia

1. Introduction

With the advancement in the areas of wireless technology and low-power electronics, a pervasive system is made possible. This system is referred to a world where computational devices are embedded in the environment for intelligent buildings and home automation, autonomous vehicles and also possible to be implanted in human bodies such as the one in body sensor networks for health monitoring. To develop a totally autonomous system, however, traditional batteries, with limited life-span have to be replaced with energy harvesters, which can provide clean and renewable electrical energy sources.

Vibration-based energy harvesting is one of the attractive solutions for powering autonomous microsystems, due to the fact that, vibration sources are ubiquitous in the ambient environment. Basically, the vibration-to-electricity conversion mechanism can be implemented by piezoelectric [1], electromagnetic [2], electrostatic [3], and magnetostrictive [4] transductions. In this thesis, piezoelectric transduction is investigated due to its high electrical output density, compatibility with conventional thick-film and thin-film fabrication technologies and ease of integration in silicon integrated circuits.

Typically, piezoelectric materials are fabricated in the form of a cantilever structure, whereby stress is induced by bending the beam configuration in an oscillating manner and generating electric charges on its electrodes, as a result of the piezoelectric effect [5]. They are widely used as sensors and actuators [6, 7]. In recent years, piezoelectric materials are advancing into another level of development whereby they are used to provide an alternative for powering wireless sensor nodes through vibrations within the environment [1, 8, 9].

Typically, the piezoelectric materials are deposited on a non-electro-active substrate such as alumina, stainless steel or aluminium. They are physically clamped at one end to a rigid base and free to move at the other end. The presence of the substrate does not contribute directly to the electrical output, but merely serves as a mechanical supporting platform, which constrain the movement on the piezoelectric materials and poses difficulties for integration with other microelectronic devices. In order to minimise the constraint, a cantilever structure, which is free from external support or attachment to a non-electro-active platform is proposed. This structure would be in free-standing form consists of only the active piezoelectric materials and electrodes, and would be able to be stressed to generate charges similar to the traditional cantilever structure.

Micro scale free-standing structures in the form of cantilever are commonly fabricated by using thick-film, thin-film and silicon micromachining technology [10]. However, thin-film and micromachining involves complex and expensive processes such as chemical vapour deposition and photolithography. Furthermore, the structures fabricated in these technologies generally are small (a few micrometers in length and width, and less than 1 μm thick) [11], therefore usually producing very low electrical output power (in order of nano-watts) and operate at high level of vibration (in order of kilohertz). The technology used for fabricating free-standing devices depends on the application, for example, in bio-molecular recognition [12], thin-film and micro-machining technologies are used to fabricate cantilevers with sub-micron dimensions. Thick-film technology is preferable to be used for fabricating bigger structures with thicknesses greater than 50 μm, and typically with area from a few mm² to a few cm², which is the size in between bulk devices and thin-film devices. Thick-film technology can be used to fill the gap between these technologies.

There are a number of challenges in the research of designing, fabricating and characterising free-standing thick-film piezoelectric cantilevers for energy harvesting. Firstly the research requires the understanding of the process conditions and limitation of thick-film technology particularly for fabricating three-dimensional structures. Thick-film technology involves processes which are hostile and destructive to ceramic free-standing structures e.g. high contact force (> 1 N) during screen-printing, high air flow curtain (> 50 l min⁻¹) in multi-zone furnace and high thick-film processing temperature (> 800 °C). The thermal expansion coefficient mismatch between electrode and piezoelectric materials could also pose a problem in fabricating straight and flat cantilever. Besides that, the mechanical properties of thick-film ceramic materials are notoriously brittle and fragile which is poor to withstand the stress induced when the structure is operated in bending mode.

The target to meet the minimum electrical energy requirement for powering the microsystem is another surmounting challenge. Typically, a ceramic cantilever structure has high mechanical Q-factor at around 150, therefore, in order to harvest maximum electrical energy, the resonant frequency of the device has to match the ambient vibration sources. The unpredictable nature of ambient vibration sources intensifies the challenges toward making thick-film free-standing structures as a useful ambient energy harvester. All of these challenges will be addressed and suggested solutions to the issues will be discussed in detail in this thesis.

2. Piezoelectricity

Piezoelectricity is the ability of certain crystals to generate a voltage when a corresponding mechanical stress is applied. The piezoelectric effect is reversible, where the shape of the piezoelectric crystals will deform proportional to externally applied voltage.

Piezoelectricity was first discovered by the brothers Pierre Curie and Jacques Curie in 1880. They predicted and demonstrated that crystalline materials like tourmaline, quartz, topaz, cane sugar, and Rochelle salt (sodium potassium tartrate tetrahydrate) can generate electrical polarization from mechanical stress. Inverse piezoelectricity was mathematically deduced from fundamental thermodynamic principles by Lippmann in 1881. Later the Curies confirmed the existence of the inverse piezoelectric effect [13].

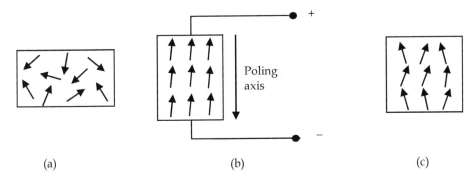

Fig. 1. Schematic diagram of the electrical domain: (a) before polarisation, (b) during polarisation and (c) after polarisation.

A piezoelectric crystal is built up by elementary cells consisted of electric dipoles, and dipoles near to each other tend to be aligned in regions called Weiss domains. These domains are randomly distributed within the material and produce a net polarisation as shown in Figure 1 (a), therefore the crystal overall is electrically neutral.

For the material to become piezoelectric, the domains must be aligned in a single direction. This alignment is performed by the poling process, where a strong field is applied across the material at the Curie temperature (a temperature above which, the piezoelectric material loss its spontaneous polarization and piezoelectric characteristics, when external electric field is not applied). The domains are forced to switch and rotate into the desired direction, aligning themselves with the applied field (Figure 1 (b)). The material is then cooled to room temperature, while the electric field is maintained. After polarisation, when the electric filed is removed, the electric dipoles stay roughly in alignment (Figure 1 (c)). Subsequently, the material has a remanent polarisation. This alignment also causes a change in the physical dimensions of the material but the volume of the piezoelectric material remains constant.

2.1 Constituent equations of piezoelectricity

One thing in common between dielectric and piezoelectrics is that both can be expressed as a relation between the intensity of the electric field E and the charge density $\mathcal{D}$. However, beside electrical properties, piezoelectric interaction also depends on mechanical properties, which can be described either by the strain, δ or the stress, σ. The relations between $\mathcal{D}_i$, E_k, δ_{ij}, and σ_{kl} can be describe in a strain-charge form of constitutive equation as,

$$\begin{bmatrix} \delta_{ij} \\ \mathcal{D}_i \end{bmatrix} = \begin{bmatrix} s_{ijkl}^E & d_{ijk} \\ d_{ikl} & \varepsilon_{ik}^T \end{bmatrix} \begin{bmatrix} \sigma_{kl} \\ E_k \end{bmatrix} \tag{1}$$

Vector $\mathcal{D}_i$ (C/m²) and E_k (N/C) are tensors of three components and the stress σ_{kl} (N/m²) and the strain δ_{ij} (m/m) are tensors of six components. d_{ikl} (C/N) is the piezoelectric charge constant and its matrix-transpose d_{ijk}, s_{ijkl}^E (m²/N) is the elastic compliance at constant electric field (denoted by the subscript E) and ε_{ik}^T (F/m) is the permittivity at constant stress (denoted by the subscript T).

The anisotropic piezoelectric properties of the ceramic are defined by a system of symbols and notations as shown in Figure 2. This is related to the orientation of the ceramic and the direction of measurements and applied stresses/forces.

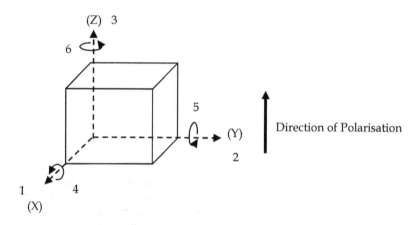

Fig. 2. Notation of piezoelectric axes.

A cantilever piezoelectric can be designed to operate in either d_{31} or d_{33} modes of vibration depending on the arrangement of the electrodes [14]. d_{31} is a thickness mode polarisation of plated electrode on the piezoelectric materials, with stress applied orthogonal to the poling direction, as shown in Figure 3 (a). d_{33} mode on the other hand, can be implemented by fabricating interdigitated (IDT) electrodes on piezoelectric materials for in-plane polarisation where stress can be applied to the poling direction, as shown in Figure 3 (b).

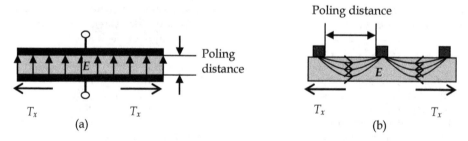

Fig. 3. Cross-sectional view of piezoelectric configuration mode, (a) d_{31} and (b) d_{33}.

2.2 Piezoelectric materials

There are a wide variety of piezoelectric materials. Some naturally exist in the form of crystals like Quartz, Rochelle salt, and Tourmaline group minerals. Some poled polycrystalline ceramics like barium titanium, and lead zirconate titanate, PZT, and polymer piezoelectric materials like polyvinylidene fluoride, PVDF and polyimide can be manufactured and easily integrated with MEMS [5].

Commercially, piezoelectric materials are manufactured in bulk form. They are fabricated from a combination of ceramic materials (in short piezoceramics) and pressed in a high temperature (1100 – 1700°C) to form a solid poly-crystalline structure. The raw material to fabricate bulk piezoelectric is in powder form. The powder is then pressed and formed into desired shapes and sizes, which is mechanically strong and dense [15]. In order to make these bulk ceramics into piezoelectric materials, electrodes are deposited onto their surface either by screen printing or vacuum deposition, and poled with electric fields of 2-8 MV·m^{-1} in an oil bath at a temperature of 130 - 220 °C [16]. Bulk piezoceramics are attractive for their high electromechanical efficiencies and high energy densities. However, bulk piezoceramics tend to be relatively thick (greater than 100 μm), which will not be sensitive and need higher energies to actuate their structures, besides that they are difficult to be processed into thickness below 100 μm, therefore limit their application in Micro-Electro-Mechanical System (MEMS). Furthermore they need to be attached to certain parts of the MEMS structures using mechanical or adhesive bonding, which is tedious and not cost effective. MEMS devices which require piezoelectric structures with features below 100 μm would usually be fabricated using thin and thick film technologies.

Piezoelectric polymer materials are attractive in fabricating flexible devices. They have much higher piezoelectric stress constants and low elastic stiffness which give them advantages in producing high sensitivity sensors compared to brittle piezoceramics. However, these materials have lower piezoelectric charge constant and are not favourable to fabricate device for electrical power generation. Polyvinylidene fluoride (PVDF) is a common piezoelectric polymer material, which was discovered by Kawai [17]. It is lightweight, tough, and can be cut to form relatively large devices. The earlier form of PVDF was in polymer sheet, which is difficult to be shaped in micro-scale and they are usually processed with a punching technique based on a micro-embossing technique which is described in the literature [18]. With the development of PVDF thin-film technology, micro-structures can be fabricated as reported by Arshak *et al* [19]. The fabrication process involved drying and curing at low temperature of around 170 °C, and was able to produce d_{33} of 24 pC/N^1 [20]. An alternative to PVDF is polyimide, a high temperature piezoelectric polymer, which can maintain its piezoelectric properties at temperature up to 150 °C as reported by Atkinson *et al* [21].

Film piezoceramics have the advantages that lie between bulk and polymer piezoelectric materials. Although film piezoceramics do not have piezoelectric activity as high as bulk piezoceramics, however, for certain applications where a device thickness has to be fabricated less than 100 μm, film piezoceramics are more favourable for their fabrication compatibility with micro scale devices. Films can be deposited directly on to a substrate, using a deposition technique that is more precise and with higher resolution. The processing temperature of film piezoceramics is in between bulk piezoceramics and piezoelectric polymers (800 °C – 1000 °C), which make it possible to be integrated with semiconductor technology. Film piezoceramics basically can be fabricated with thin- and thick-film technologies. Thin-film technologies involve physical vapour deposition, chemical vapour deposition, and solution deposition, which fabricate films with typical thickness less than 5 μm. For thicker films (10 μm – 100 μm), thick-film technology is preferable. The technology involves a screen printing method, where each layer of ceramic thick film will be printed on a substrate followed by drying and curing processes.

2.3 Lead zirconate titanate (PZT)

Research and development in high performance piezoelectric ceramic had attracted great attention since the discovery of barium titanium oxide in 1940 [22]. This was followed by the discovery of lead titanate zirconate (PZT) in 1950s by Bernard Jaffe [23]. Compared to barium titanium oxide, PZT has a higher Curie point, higher total electric charge, and higher coercive voltage. PZT can be processed in bulk, thin-film, thick-film, and polymer forms in applications suited to their individual characteristics.

Thick-film PZT materials can be classified as 'hard' and 'soft', according to their coercive field during field-induced-strain actuation and Curie temperature [24]. A 'hard' piezoceramic has larger coercive field (greater than 1 kV/mm) and higher Curie point (T_C > 250 °C) compared to 'soft' piezoceramic, which has moderate coercive field (between 0.1 and 1 kV/mm) and moderate Curie point (150 °C < T_C < 250 °C) . Examples of 'hard' PZTs are Pz26 from Ferroperm Piezoceramics [16] and PZT-401 from Morgan Electroceramics [25]. Their typical applications are high power ultrasonics for cleaning, welding and drilling devices. Their distinctive characteristics include high mechanical factor, high coercive field, and low dielectric constant, which make them capable to be used in underwater applications and high voltage generators.

Compared to its counterpart, 'Soft' PZTs have lower mechanical Q-factor, higher electromechanical coupling coefficient, and higher dielectric constant, which are useful to fabricate sensitive receivers and applications requiring fine movement control, for instant in hydrophones and ink jet printers. Other applications ranging from combined resonant transducers (for medical and flow measurements) to accelerometer and pressure sensors [26]. Examples of soft PZTs are Pz27 and Pz29 from Ferroperm Piezoceramics. Pz27 and Pz29 have similar properties as PZT-5A and PZT-5H respectively from Morgan Electroceramics [25].

3. Vibration energy harvesting

Piezoelectric is one of the four general types of mechanical-to-electrical energy conversion mechanisms for harvesting vibration energy [27]. The other three are electromagnetic [2], electrostatic [3] and magnetostrictive [4]. With the improvement of piezoelectric activity, the PZT piezoelectric materials (traditionally used to fabricate sensing devices) are becoming popular in fabricating micro-power generators for the application of embedded and remote systems [28]. Micro-generator is the term often used to describe a device which produces electrical power in micro-Watt scale, while energy harvester is a more general term for describing a device which produces power derived from external ambient sources (e.g. solar, vibration, thermal and wind energy). Both of these terms will be used interchangeably in this thesis where appropriate.

The vibration energy harvesting of piezoelectric materials is based on the concept of shunt damping to control mechanical vibration [29], however, rather than dissipating the energy through joule heating, the energy is used to power some electronic devices.

In order to estimate the output power from a vibration energy harvester, analytical models have been developed over the years. A generic energy conversion model followed by a specific conversion model for piezoelectric will be discussed in the following section.

3.1 Generic mechanical-to-electrical conversion model

One of the earliest general models for energy harvesters was proposed by William and Yates [30]. The model is represented as a single-degree-of-freedom linear mass-spring-damper system as illustrated in Figure 4

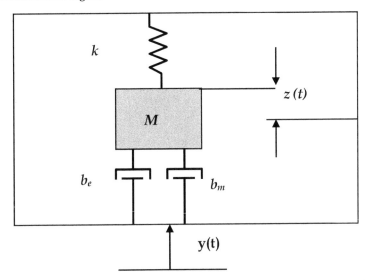

Fig. 4. A schematic diagram of a spring-mass-damper system of a piezoelectric FSD, based on the model developed by Williams et al [30].

When the system with lump mass, M is excited with a displacement of $y(t)$ relative to the system housing, a net displacement $z(t)$ is produced and the generic equation derived from Newton's second law can be written as in equation (2), with the assumption that the source of the vibration is unlimited and unaffected by the system. The general single degree of freedom model can be written as,

$$M\ddot{z}(t) + \left(b_e + b_m\right)\dot{z}(t) + \kappa z(t) = -M\ddot{y}(t) \tag{2}$$

where κ is the spring constant. For a piezoelectric device, the damping effect of the system is related to its induced damping coefficient, b (with subscripts e and m referring to electrical and mechanical damping respectively), which can be written in relation to damping ratios, ζ and undamped natural frequency, ω_n as,

$$b_{e,m} = 2M\omega_n\zeta_{e,m} \tag{3}$$

As the system undergoes harmonic motion relative to the base with external excited displacement $y(t) = Y\sin(\omega t)$, there is a net transfer of mechanical power into electrical power. By solving the equation (2) and $P = \frac{1}{2}b_e\dot{z}$ (electrical induced power), the magnitude of the generated electrical power can be written as,

$$P = \frac{M\zeta_e\omega^3 \left(\dfrac{\omega}{\omega_n}\right)^3 Y^2}{\left[2\zeta_T\left(\dfrac{\omega}{\omega_n}\right)\right]^2 + \left[1-\left(\dfrac{\omega}{\omega_n}\right)^2\right]^2} \tag{4}$$

where ζ_T is the total damping ratio ($\zeta_T = \zeta_e + \zeta_m$), and ω is the base excited angular frequency and Y is the amplitude of vibration. When the device is operated at its resonant frequency ω_n, maximum power can be produced and equation (4) is simplified to,

$$P_{max} = \frac{M\zeta_e a_{in}^2}{4\omega_n\left(\zeta_e + \zeta_m\right)^2} \tag{5}$$

where a_{in} is input acceleration from vibration source ($a_{in} = \omega_n^2 Y$). This equation shows that input acceleration is the major factor for increasing the output power from the piezoelectric FSDs. By maintaining the frequency of the vibration source to match the natural frequency of the device, the electric power generated by the device is proportional to the square of the source acceleration.

3.2 Analytical model of piezoelectric harvester

Although the mass-spring-damper system with lumped parameters is more suitable to represent a simple electromagnetic vibration-to-electric energy conversion model, it gives an insight of a general mechanism of mechanical to electrical transduction model which include piezoelectric transduction.

A more specific piezoelectric energy harvester model, where the mechanism of piezoelectric transduction due to the constitutive relations according to equation (1) is taken into account, has been proposed by duToit et al [31], with an additional term related to undamped natural frequency, ω_n, piezoelectric charge constant, d_{33} and output voltage, v being added to the single-degree-of-freedom equation (2). However, the model does not give a clear picture of optimum load resistance at resonant frequency. An improved model by Roundy et al [8] suggested an analogous transformer model representing the electromechanical coupling, while the mechanical and the electrical domains of the piezoelectric system are modelled as circuit elements, as shown in Figure 5.

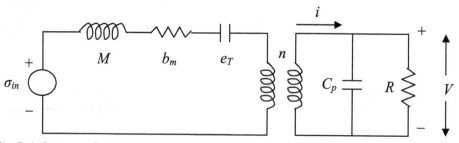

Fig. 5. A diagram of an analogous circuit for a piezoelectric vibrated device with a resistive load.

The mechanical domain of the equivalent circuit consists of inductor, resistor and capacitor which represents the mass of the generator, M, the mechanical damping, b_m, and mechanical stiffness, e_T respectively. At the electrical domain, C_p is the capacitance of the piezoelectric and R is the external resistive load, while n is the equivalent turn ratio of the transformer which is proportional to the piezoelectric charge constant d_{31}. V is the voltage across the piezoelectric and i is the current flow into the circuit, which are analogues to the stress and the strain rate respectively. The output voltage at resonant frequency derived from the model is,

$$V = \frac{-j\omega \dfrac{e_p d_{31} h_p B}{\varepsilon} a_{in}}{\left[\dfrac{1}{RC_p}\omega_r^2 - \left(\dfrac{1}{RC_p} + 2\zeta_T\omega_r\right)\omega^2\right] + j\omega\left[\omega_r^2\left(1 + k^2\right) + \dfrac{2\zeta_T\omega_r}{RC_p} - \omega^2\right]} \tag{6}$$

where j is the imaginary number, ω is the driving frequency (Hz), ω_r is the fundamental resonant frequency of the cantilever (Hz), E_T is the elastic constant for the composite structure (N/m²), d_{31} is the piezoelectric charge coefficient (C/N), h_P is the thickness of the piezoelectric material, ε is the dielectric constant of the piezoelectric material (F), B is a constant related to the distance from the piezoelectric layer to the neutral axis of the structure, ζ_T is the total damping ratio, k_{31} is the piezoelectric coupling factor and C_P is the capacitance of the piezoelectric material. The root mean square (rms) power is given as $|V|^2/2R$, therefore from equation (6), the rms value of power transferred to the resistive load can be written as,

$$P = \frac{1}{\omega_r^2} \frac{RC_P^2\left(\dfrac{e_t d_{31} h_P B}{\varepsilon}\right)^2 a_{in}^2}{\left(4\zeta_T^2 + k_{31}^4\right)\left(RC_P\omega_r\right)^2 + 4\zeta_T k_{31}^2\left(RC_P\omega_r\right) + 4\zeta_T^2} \tag{7}$$

More complex models have been developed by Erturk and Inman [32, 33]. Instead of a single-degree-of-freedom model, they had developed a distributed parameter electromechanical model which incorporates Euler-Bernoulli beam theory with the piezoelectric constitutive equation. The detail of this model will not be discussed in this research work. However, both models agree to a certain extent that at resonant frequency, the output power is proportional to the square power of the piezoelectric charge coefficient, the elasticity of the cantilever, the thickness of the piezoelectric material and the effective mass of the cantilever, all but the first of which are controllable by design. It is also found that the input acceleration from base excitation, a_{in} ($a_{in} = \omega_n^2 Y$) plays an important part in output power generation. However, for the application of energy harvesters, the acceleration level from an ambient vibration source is a natural phenomenon, which is not controllable. Therefore the energy harvester has to be designed to suit the specific application, though the model gives a good estimation for the potential power generation.

3.3 Cantilever-based piezoelectric energy harvesters

The most common piezoelectric energy harvesters are in the form of a cantilever, due to its simple geometry design and relative ease of fabrication. The structures usually consist of a strong flexible supporting platform with one end fixed to the base on the substrate. Piezoelectric materials are deposited on either one side (unimorph) or both sides (bimorph) of the platform with the intention to strain the piezoelectric films and generate charges from the piezoelectric d_{31} effect. This bending mode operation is effectively generating electrical energy when they are exposed to continuous harmonic vibration sources.

The flexible supporting platform is not electrically active but acts as a mechanical support to the whole structure. It can be stainless steel, aluminium plate or micromachined silicon depending on the fabrication process and the scale of the device. One of the earliest examples using stainless steel as the supporting platform was developed by Glynne-Jones *et al* [1]. They developed a cantilever with a tapered profile as shown in Figure 6, in order to produce constant strain in the piezoelectric film along its length for a given displacement. The generator was fabricated by screen-printing a layer of PZT-5H with a thickness of 70 μm on both sides of a stainless steel beam with length 23 mm and thickness 100 μm to form a bimorph cantilever. The device was found to operate at its resonant frequency of 80.1 Hz and produced up to 3 μW of power when driving an optimum resistive load of 333 kΩ.

Another example using stainless steel as the centre supporting platform was developed by Roundy *et al* [34]. Instead of a tapered profile, they simplified their model into a rectangular cantilever with constant width. Based on the model, a prototype micro-generator was fabricated in a form of bimorph structure which consisted of two sheets of PZT attached to both sides of a steel centre shim. The structure with total size of about 1 cm^3 included a proof mass attached at the tip of the cantilever as shown in Figure 7 (a) was excited at 100 Hz with an acceleration magnitude of 2.25 m/s^2. A maximum output power of about 70 μW was measured when driving a resistive load of about 200 kΩ. An improved version of the prototype was developed with a cantilever with total length of 28 mm, width 3.2 mm and PZT thickness of 0.28 mm, attached with proof mass of length 17 mm, width 3.6 mm and height 7.7 mm as shown in Figure 7 (b), produced a maximum power of 375 μW when excited to its resonant frequency of 120 Hz at an acceleration of 2.5 m/s^{-2} [8].

An example of micromachined silicon MEMS cantilever has been developed by Jeon *et al* [11], as shown in Figure 8. The cantilever was fabricated by depositing a membrane layer of silicon oxide, a layer of zirconium dioxide which acts as a buffer layer, sol-gel deposited PZT layer and a top interdigitated Pt/Ti electrode on silicon substrate. A proof mass can be added to the cantilever by spin-coating and patterned with a layer of SU-8 photoresist. The beam is releasing by undercutting the silicon substrate using a vapour etching process. The cantilever with a dimension of 170 μm x 260 μm was found to have a fundamental resonant frequency of 13.9 kHz, which was able to generate an electrical power of 1 μW at a base displacement of 14 nm when driving a resistive load of 5.2 MΩ.

In another study, Sodano *et al* [35] compared the efficiencies of three piezoelectric materials: PZT, Quick Pack (QP) actuator and Macro-Fiber Composite (MFC) as shown in Figure 9. The PZT material was PSI-5H4E piezoceramic obtained from Piezo System Inc with a length of 63.5 mm and width 60.32 mm. The QP actuator is a bimorph piezoelectric device developed by Mide Technology Corporation, with length 101.6 mm and width 25.4 mm. It was fabricated from a monolithic piezoceramic material embedded in an epoxy matrix, which is ready to be clamped at one end to form a cantilever. The MFC prototype was developed by NASA, consists of thin PZT fibres embedded in a Kapton film with length

82.55 mm and width 57.15 mm and connected with an interdigitated electrode (IDE) pattern. Both the brittle PZT material and the flexible MFC were bonded on a 0.0025 in. aluminium plate and clamped at one end. From their experiment, they found that the PZT performed better than the other two prototypes, with an efficiency of 4.5 % compared to 1.75 % for MFC prototype at resonant frequency. However, their research interest was at the time aimed at recharging nickel metal hydride batteries and they did not report on the maximum output power of the prototypes.

As according to equation (1.6), the output voltage is proportional to the distance from the piezoelectric to the neutral axis of the structure, therefore it is desirable to fabricate thicker cantilever structures in generating more electrical energy. Thicker structures, however, are less elastic and not suitable for harvesting energy from ambient vibration sources. Alternatively, Wang *et al* [36] made an improvement to the cantilever structure by separating two plates of PZT to form an air-spaced cantilever as shown in Figure 10, which increases the distance between the piezoelectric layer and the neutral plane thus increasing the output voltage generation. The two PZT plates were formed by adhering PZT sheets (Piezo System, Inc) with thickness of 127 μm on both sides of an aluminium plate. Both of the PZT plates with length 7 mm were separated at 221 μm from its middle plane to the neutral plane and attached with proof mass with dimension 16 x 9.2 x 0.31 mm. The device was tested with a speaker with a consistent sinusoidal signal maintained with commercial accelerometer. An output of 32 mV/g was measured at its resonant frequency of 545 Hz.

Another issue faced by cantilever structures in harvesting energy is the movement constraint of piezoelectric materials when the structures are deformed. As the piezoelectric materials are rigidly clamped on substrate, therefore the stress imposed on the active materials is dependent on the elasticity of the substrate which prohibits the materials to perform at their best. One of the solutions to get rid of the constraint is by fabricating the piezoelectric materials on an elevated structure without substrate in a free-standing fashion.

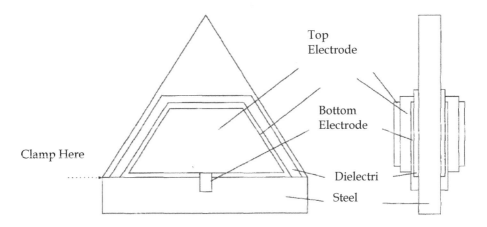

Fig. 6. Design of prototype generator (after [1]).

Fig. 7. A A rectangular cantilever microgenerator prototype (b) An improved version (after [8]).

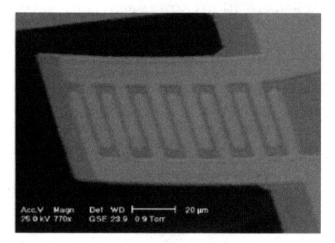

Fig. 8. MEMS micromachined IDE pattern cantilever (after [11]).

3.4 Free-standing cantilever structures

Figure 11 shows free-standing cantilever structures fabricated using thick-film technology [37]. The adopted fabrication technique results in the formation of a stand-alone structure by a process of burning out a sacrificial layer at elevated temperatures. The structure is one that stands on its own foundation and is free from external support or attachment to a non-electro-active platform. Besides eliminating the constraint imposed by the substrate on the piezoelectric materials, other advantages of free-standing thick-film structures is their ability to provide a support structure upon which other sensing materials can be deposited and ease of integration with electronic circuits. Such structures are three-dimensional micromechanical structures and are analogous to silicon micro-machined MEMS. The main difference is that free-standing thick-film structures are formed without the need for supporting platforms, which are passive mechanical elements that do not directly contribute to the generation of electrical energy. It is therefore desirable for them to be thin and flexible.

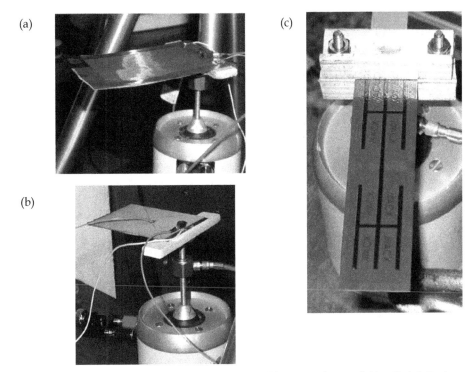

Fig. 9. Cantilever configuration of a: (a) MFC plate, (b) a PZT plate and (c) a Quick Pack actuator (after [35]).

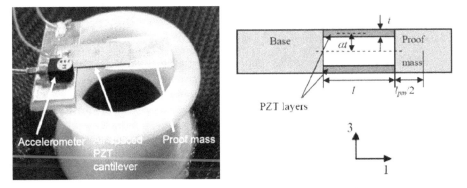

Fig. 10. Schematic structure of the vibration energy harvester based on air-spaced piezoelectric cantilevers (after [36]).

4. Thick-film technology

Thick-film technology is distinguished from other fabrication technologies by the sequential processes of screen-printing, drying and firing (curing). Screen-printing is possibly one of the oldest forms of graphic art reproduction and traditionally silk screen printing was used to transfer patterns to printable surface such as clothes, ceramics, glass, polyethylene and metals [6].

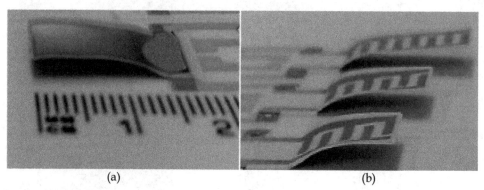

(a)   (b)

Fig. 11. Thick-Film Free-standing structures with: (a) plated electrodes and (b) interdigitated (IDT) electrodes.

The process is ideal for mass production with the ability to produce films of 10-50 μm thick in one print whilst other deposition and printing techniques require many hours of processing to achieve the same thickness. Limitations of conventional screen-printing are feature size and geometry with a minimum line width and separation distance around 100-150 μm.

4.1 Evolution of thick-film technology

Thick-film technology is traditionally used to manufacture resistor networks, hybrid integrated circuits, and other electronic components [38]. In the past two decades, research in thick-films has been extended to include sensing capabilities [39]. One of the prominent applications of thick-film as a sensing element was as a strain gauge [40, 41].

One of the earliest piezoelectric devices fabricated with thick-film technology was reported by Baudry in 1987 [42]. Following on from the discovery of high piezoelectric activity materials such as lead zirconate titanate (PZT) brought thick-film technology to another level of development, where it is possible to fabricate micro-generators for embedded and remote systems [43].

Thick-film micro-generators are commonly fabricated in the form of a cantilever to harvest energy from bending mode as discussed in the previous section. Another example of a thick-film generator is based on the thermoelectric principle. This type of generator can be fabricated from high Seebeck coefficient materials such as bismuth telluride, which has the potential to convert body temperature changes into useful electric power sources [44]. However, the development of the generator is still in an early stage to investigate the feasibility for implantable biomedical applications.

There are many other interesting applications which need acceptable acoustic outputs for instance in micro-fluidic application for carrying out chemical and biological analysis, which is known as micro total analysis systems (μTAS) or "Lab on a Chip" [45]. Thick-film technology was used in fabricating multi-layered resonators for use as a micro-fluidic filter to separate particles within the fluid by ultrasonic standing waves.

4.2 Standard thick-film fabrication process

Piezoelectric paste is the main component in thick-film technology. It is a composite of finely powdered piezoelectric ceramic dispersed in a matrix of epoxy resin which was applied as a film onto a substrate by scraping with a blade. Alternatively, thick-film piezoelectric materials can be made into a form of water-based paint as described by Hale [46]. The piezoelectric paint consists of polymer matrix to bind PZT powder and cured at ambient temperature. One of the advantages of this paint is able to spray on flexible substrate materials and has found application in dynamic strain sensors.

The basic equipment used for processing screen-printed thick-film are the screen, screen printer, infrared dryer and multi-zone furnace. A typical thick-film screen is made from a finely woven mesh of stainless steel, polyester or nylon. For optimum accuracy of registration and high resolution device printing, a stainless steel screen is preferred. The screen is installed in a screen printer, which is necessary for accurate and repeatable printing. The screen printer consists of a squeegee, screen holder and substrate work-holder. Before a printing process is started, the gap between substrate and screen is adjusted to be around 0.5 mm to 1 mm, depending on the screen material and the resolution required for the print (a bigger gap is necessary for flexible materials such as polyester screen, and also as a requirement for higher definition printing).

The substrate work-holder is aligned according to the printing pattern on the screen. Once the setting is correct, a printable material (paste / ink) is then smeared across the pattern on the screen as shown in Figure 12 (a). A squeegee is then brought in contact with the screen with applied force, which deflects the screen (Figure 12 (b)) and the paste is drawn through by surface tension between the ink and substrate and deposits on the substrate under the screen which is rigidly held by the substrate holder as shown in Figure 12 (c).

After screen-printing, an irregular surface pattern caused by the screen mesh appears on the wet print surface. Therefore before the drying process, the printed layer needs to be left to settle for about 10 minutes otherwise a uniform device thickness will not be achieved. The drying process is carried out in an infra-red belt conveyor or a conventional box oven at a temperature around 150 °C for 10–15 minutes. The function of the drying process is to remove the organic solvents by evaporation from the wet print and retain a rigid pattern of films on the substrate. Normally, the thickness of the film will be reduced by up to half of its original printed thickness after the drying process. A thicker film can be formed by printing another layer of film directly onto the dried film. The next stage of the process is co-firing, where the dried films are annealed in a multi-zone belt furnace. This is to solidify the composite of the films which consist of glass frit and active particles (e.g. PZT). During the process, the glass melts and binds the active particles together and adheres to the substrate.

The main concerns for piezoelectric thick-film fabrication are to produce films that are uniform in thickness, crack-free, have high mechanical density, are reproducible, and with high piezoelectric performance. Reproducible and high piezoelectric performance can be achieved by formulating correct paste composition. The curing or co-firing temperature is crucial as well to determine piezoelectric properties of the films, while screen-printing with correct squeeze pressure and snap height can control the film thickness and uniformity. Screen mesh and emulsion thickness are also important to determine deposition resolution and quality of prints.

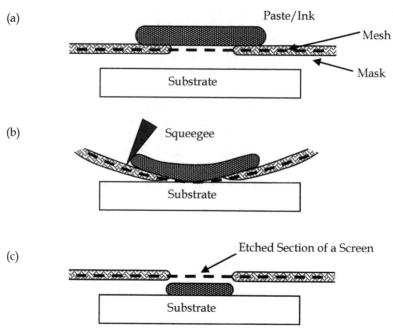

Fig. 12. Thick-film screen printing steps.

4.3 Thick-film vs thin-film fabrication technologies

Thin-film is defined as a layer of materials having thickness from a fraction of nanometer to several micrometers. On the other hand, thick-film literally is thicker than the former, with thickness up to several hundred micrometers. However, a more general term to distinguish both of them is the fabrication technologies involved in forming the films.

Thin-film fabrication technologies are well established and have been used for decades in semiconductor fabrication industries. Thin-film technologies involve physical vapour deposition, chemical vapour deposition, and solution deposition. One of the techniques in fabricating free-standing structure is surface micromachining. Surface micromachined features are built up, layer by layer on a surface of a substrate. Usually sacrificial layer techniques are used where the active layers which are the eventual moving structures are deposited on temporary rigid platforms. The platforms will then be removed, usually by etching away the materials. These platforms are called 'sacrificial layers', since they are 'sacrificed' to release the materials above them. Unlike bulk micromachining, where a silicon substrate is selectively etched to produce free-standing structures, surface micromachining is based on the deposition and etching of different structural layers on top of the substrate. Therefore the substrate's properties are not critical. Expensive silicon wafer can be replaced with cheaper substrates, such as glass, and the size of the substrates can be much larger compared to those used in bulk micromachining. The sacrificial layer for surface micromachining could be silicon oxide, phosphosilicate glass or photoresist. Figure 13 shows the fabrication steps of surface micromachining in building a free-standing structure.

Thick-film free-standing structure can be fabricated by using sacrificial layer techniques as those used in the conventional thin-film processing technologies as described above. One of the examples of fabrication incorporating sacrificial layer techniques is polymer free-standing structures based on SU-8 [47]. The structures were fabricated using Cu and lift-off resist as the sacrificial layers, where they were wet-etched at the final stage of the process. Piezoelectric polymer free-standing structures were fabricated by Atkinson et al [21], using piezoelectric polyimide as the active material and photoresist as the sacrificial layer. The process was based on conventional lithography and metallization techniques and the fabrication steps are shown in Figure 14.

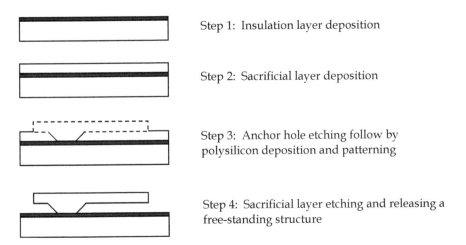

Step 1: Insulation layer deposition

Step 2: Sacrificial layer deposition

Step 3: Anchor hole etching follow by polysilicon deposition and patterning

Step 4: Sacrificial layer etching and releasing a free-standing structure

Fig. 13. Fabrication steps of surface micromachining based on sacrificial layer technique [10].

Stecher [48] developed a thick-film free-standing structure by combining the processing of air and nitrogen fireable materials on the same substrate, where initially a carbon-like filler was printed and dried on those areas of the substrate for the structure to be free supporting at a later stage. The filler has to prevent the successively printed dielectric from being bonded to the substrate. This was followed by a second step where the dielectric material is printed on top of the filler and parts of the substrate, where the part that printed on the substrate will form a rigid base to support the free-standing structure.

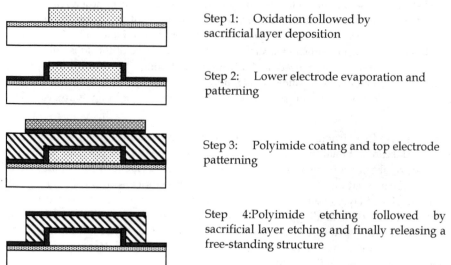

Step 1:   Oxidation followed by sacrificial layer deposition

Step 2:   Lower electrode evaporation and patterning

Step 3:   Polyimide coating and top electrode patterning

Step 4:Polyimide etching followed by sacrificial layer etching and finally releasing a free-standing structure

Fig. 14. Piezoelectric polyimide free-standing structure fabrication steps [21].

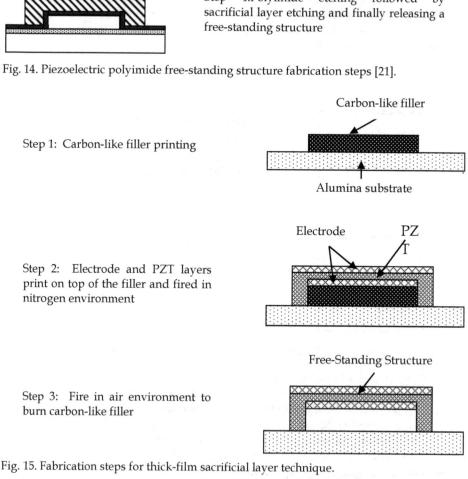

Step 1: Carbon-like filler printing

Step 2:   Electrode and PZT layers print on top of the filler and fired in nitrogen environment

Step 3:   Fire in air environment to burn carbon-like filler

Fig. 15. Fabrication steps for thick-film sacrificial layer technique.

The dried paste is then co-fired in a nitrogen atmosphere. The nitrogen must be used because the filler must not be burnt out before the glass-ceramic has sintered. The process is repeated to form a multilayer composite film. Finally, the composite film is co-fired in an air environment, where the carbon filler acting as a sacrificial layer is burnt out without residues, releasing a composite thick-film free-standing structure. The fabrication steps are shown in Figure 15. Similarly, the proposed free-standing energy harvester structures as described in section 3.4 can be fabricated using this technique.

5. Energy harvesters performance comparison

Over the years, many micro-generator prototypes have been fabricated. The most common vibration energy harvester is based on an electromagnetic principle because at present, the output powers produced by electromagnetic generators are greater than piezoelectric and electrostatic based generators. However, with recent improvement in piezoelectric activity in PZT and the ability to be incorporated within simple cantilever structures, which is relatively easy to be fabricated and integrated with microelectronic systems, piezoelectric methods are an attractive alternative for future investigation.

Each of the energy harvesters was being claimed to demonstrate better performance in one way than another. The most common comparison merit is the electrical output power density. Although power density comparison can give an idea of the performance of an energy harvester, it does not explain the influence of the excitation source. As according to Equation (5), the output power of a resonant device is closely dependent on the amplitude of an excitation source. However, to make the comparison meaningful, all the energy harvesters have to be excited at a fixed vibration characteristic (e.g. adjust acceleration level at resonant frequency of the tested devices to give a fixed vibration amplitude), which is impossible as the size of the energy harvesters range from micro to centimetre scales depending on the fabrication technology. Micro-scale devices are more sensitive to micro-scale vibration amplitudes (a few nano- to micrometer), while centimetre scale devices do not show their optimum performances if excited at these same levels, therefore it is not appropriate to make a comparison in terms of power density.

There are other alternative ways to compare the energy harvesters in a more universal metric, for example, a normalised power density (NPD) suggested by Beeby et al [2], in which the power density is divided by the source acceleration amplitude squared. Volume figure of merit, FoM_V, suggested by Mitcheson et al [49], measures the performance as a percentage comparison to its maximum possible output for a particular device. The maximum possible output is proportional to the resonant frequency of the device to the power of three and the overall size of a device with an assumption that the device (with a proof mass) has the density of gold, occupying half of the total volume and the other half is room for displacement,

$$FoM_V = \frac{\text{Measured Power Output}}{\frac{1}{16}Y_0\rho_{AU}Vol^{4/3}\omega^3} \tag{8}$$

A few recently published experimental results of fabricated energy harvesters are listed and summarised in Table 2.1. The table is divided into three sections according to the mechanism of power conversion. Each of the micro-generator is identified by the first author and the year of the publication.

Micro-generator	Power (µW)	Freq (Hz)	Volume* (cm³)	Input Acceln (m/s²)	NPD (kgs/m³)	FoMv (%)
Piezoelectric						
Glynne-Jones, 2000 [50]	3	80.1	70	NA	NA	NA
Roundy, 2003 [51]	375	120	1.0	2.5	60	1.65
Tanaka, 2005 [52]	180	50	9	1	20.5	0.26
Jeon, 2005 [11]	1.0	1.4×10^4	2.7×10^{-5}	106.8	3.2	1.10
Fang, 2006 [20]	2.16	609	6.0×10^{-4}	64.4	0.9	1.44
Reilly, 2006 [53]	700	40	4.8	2.3	28.2	1.25
Lefeuvre, 2006 [54]	3.0×10^5	56	34	0.8	1.42×10^4	81.36
Ferrari, 2006 [55]	0.27	41	0.188	8.8	0.018	0.01
Mide, 2010 [56]	8.0×10^3	50	40.5	9.8	2.1	0.16
Kok, 2011 [37]	110	155	0.12	4.9	38.2	3.23
Electromagnetic						
Ching, 2000 [57]	5	104	1	81.2	7.6×10^{-4}	7.82×10^{-4}
Li, 2000 [58]	10	64	1.24	16.2	0.03	0.01
Williams, 2001 [59]	0.33	4.4×10^3	0.02	382.2	1.1×10^{-4}	4.8×10^{-5}
Glynne-Jones, 2001 [60]	5.0×10^3	99	4.08	6.9	26.1	1.49
Mizuno, 2003 [61]	4.0×10^{-4}	700	2.1	12.4	1.24×10^{-6}	2.26×10^{-8}
Huang, 2007 [62]	1.44	100	0.04	19.7	0.09	0.07
Beeby, 2007 [2]	46	52	0.15	0.6	884	24.8
Torah, 2008 [63]	58	50	0.16	0.6	1.0×10^3	29.4
Ferro Solution, 2008 [64]	1.08×10^4	60	133	1	84.4	0.36
Perpetuum, 2009 [65]	9.2×10^4	22	130.7	9.8	7.33	0.85
Electrostatic						
Tashiro, 2002 [66]	36	6	15	12.8	0.015	0.017
Mizuno, 2003 [61]	7.4×10^{-6}	743	0.6	14.0	6.34×10^{-8}	1.86×10^{-9}
Arakawa, 2004 [67]	6.0	10	0.4	4.0	0.96	0.68
Despesse, 2005 [68]	1.0×10^3	50	18	8.8	0.7	0.06
Miao, 2006 [69]	2.4	20	0.6	2.2×10^3	8.0×10^{-7}	0.02
Basset, 2009 [70]	0.06	250	0.07	2.5	0.15	4.9×10^{-3}

* Device size does not include the electrical possessing and storage circuits
NA = Data is not available from literature

Table 1. Comparison of a few key experimental energy harvesters.

6. References

[1] Glynne-Jones, P., S.P. Beeby, and N.M. White, *Towards a piezoelectric vibration-powered microgenerator.* IEE Science Measurement and Technlogy, 2001. 148(2): p. 68-72.

[2] Beeby, S.P., Torah, R. N., Tudor, M. J., Glynne-Jones, P., O'Donnell, T., C.R. Saha, and S. Roy, *A micro electromagnetic generator for vibration energy harvesting.* Journal of Micromechanics and Microengineering, 2007. 17(7): p. 1257-1265.

[3] Mitcheson, P.D., Miao, P., Stark, B. H., Yeatman, E. M., A.S. Holmes, and T.C. Green, *MEMS electrostatic micropower generator for low frequency operation.* Sensors and Actuators A: Physical, 2004. 115(2-3): p. 523-529.

[4] Wang, L. and F.G. Yuan, *Vibration energy harvesting by magnetostrictive material.* Smart Materials and Structures, 2008. 17(4): p. 045009.

[5] Polla, D.L. and L.F. Francis, *Processing and characterization of piezoelectric materials and integration into microelectromechanical systems.* Annual Review of Materials Science, 1998. 28: p. 563-597.

[6] White, N.M. and J.D. Turner, *Thick-film sensors: past, present and future.* Meas.Sci.Technol., 1997. 8: p. 1-20.

[7] Hoffmann, M., Kuppers, H., Schneller, T., Bottger, U., Schnakenberg, U., W. Mokwa, and R. Waser. *A new concept and first development results of a PZT thin film actuator A new concept and first development results of a PZT thin film actuator.* in *Applications of Ferroelectrics, 2000. ISAF 2000. Proceedings of the 2000 12th IEEE International Symposium on.* 2000.

[8] Roundy, S. and P.K. Wright, *A piezoelectric vibration based generator for wireless electronics.* Smart Materials and Structures, IOP, 2004. 12: p. 1131-1142.

[9] Sodano, H.A., G. Park, and D.J. Inman, *Estimation of electric charge output for piezoelectric energy harvesting.* Strain, 2004. 40: p. 49-58.

[10] Yalcinkaya, F. and E.T. Powner, *Intelligent Structures.* Sensor Review, 1996. 16: p. 32-37.

[11] Jeon, Y.B., Sood, R., J.h. Jeong, and S.G. Kim, *MEMS Power Generator with Transverse Mode Thin Film PZT.* Sensor and Actuators A, 2005(122): p. 16-22.

[12] Fritz, J., et al., *Translating Biomolecular Recognition into Nanomechanics.* Science, 2000. 288(5464): p. 316-318.

[13] Mason, W.P., *Piezoelectric crystals and their application to ultrasonics.* 1950: D. Van Nostrand Company Inc.

[14] Bernstein, J.J., Bottari, J., Houston, K., Kirkos, G., Miller, R., Xu, B., Y. Ye, and L.E. Cross. *Advanced MEMS ferroelectric ultrasound 2D arrays.* in *Ultrasonics Symposium, 1999. Proceedings. 1999 IEEE.*

[15] Jordan, T.L. and Z. Ounaies, *Piezoelectric Ceramics Characterization.* 2001, NASA/CR-2001-211225 ICASE Report, No. 2001-28.

[16] *High Quality Components and Materials for The Electronic Industry.* 2003, Ferroperm Piezoceramics.

[17] Kawai, H., *The piezoelectricity of poly(vinylidene fluoride).* Jpn.J.Appl.Phys., 1969. 8: p. 975.

[18] Fu, Y., Harvey, Erol C., M.K. Ghantasala, and G.M. Spinks, *Design, fabrication and testing of piezoelectric polymer PVDF microactuators.* Smart Materials and Structures, 2006. 15(1): p. S141-S146.

[19] Arshak, K.I., D. McDonough, and M.A. Duncan, *Development of new capacitive strain sensors based on thick film polymer and cermet technologies.* Sensors and Actuators A., 2000. 79: p. 102-114.

[20] Fang, H.B., Liu, Jing Quan, Xu, Zheng Yi, Dong, Lu, Wang, Li, Chen, Di, B.C. Cai, and Y. Liu, *Fabrication and performance of MEMS-based piezoelectric power generator for vibration energy harvesting.* Microelectronics Journal, 2006. 37(11): p. 1280-1284.

[21] Atkinson, G.M., Pearson, R. E., Ounaies, Z., Park, C., Harrison, J. S., W.C. Wilson, and J.A. Midkiff, *Piezoelectric polyimide MEMS process.* NASA 11th Symposium: May 28 - 29, 2003.

[22] Jaffe, H., *Piezoelectric Ceramics.* Journal of The American Ceramic Society, 1958. 41(11): p. 494-498.

[23] Jaffe, B., Cook Jr, W. R., Jaffe, H., J.P. Roberts, and P. Popper, *Piezoelectric Ceramics*. 1971, London: Academic Press Inc.

[24] Giurgiutiu, V. and S.E. Lyshevski, *Electroactive and Magnetoactive Materials*, in *Micromechatronics: modeling, analysis, and design with Matlab*. 2003, CRC Press. p. 357-415.

[25] *Piezoelectric ceramics data book for designers*. 1999, Morgan Electroceramics.

[26] Van Lintel, H.T.G., F.C.M. Van De Pol, and S. Bouwstra, *A piezoelectric micropump based on micromachining of silicon*. Sensors and Actuators, 1988. 15(2): p. 153-167.

[27] Erturk, A., J. Hoffmann, and D.J. Inman, *A piezomagnetoelastic structure for broadband vibration energy harvesting*. Applied Physics Letters, 2009. 94(25): p. 254102-3.

[28] Anton, S.R. and H.A. Sodano, *A review of power harvesting using piezoelectric materials (2003–2006)*. Smart Materials and Structures, 2007. 16(3): p. R1-R21.

[29] Liao, Y. and A. Sodano, *Optimal parameters and power characteristics of piezoelectric energy harvesters with an RC circuit*. Smart Mater.Struct., 2009. 18(045011).

[30] Williams, C.B. and R.B. Yates, *Analysis of a micro-electric generator for microsystems*. Transducers 95/Eurosensors IX, 1995: p. 369-372.

[31] duToit, N.E., B.L. Wardle, and S.G. Kim, *Design Considerations for MEMS-Scale Piezoelectric Mechanical Vibration Energy Harvesters*. Integrated Ferroelectrics, 2005(71): p. 121-160.

[32] Erturk, A. and D.J. Inman, *A Distributed Parameter Electromechanical Model for Cantilevered Piezoelectric Energy Harvesters*. Journal of Vibration and Acoustics, 2008. 130(4): p. 041002-041015.

[33] Erturk, A. and D.J. Inman, *Issues in mathematical modeling of piezoelectric energy harvesters*. Smart Materials and Structures, 2008. 17(6): p. 065016.

[34] Roundy, S., P.K. Wright, and J. Rabaey, *A study of low level vibrations as a power source for wireless sensor nodes*. Computer Communications, 2003. 26: p. 1131-1144.

[35] Sodano, H.A., D.J. Inman, and G. Park, *Comparison of piezoelectric energy harvesting devices for recharging batteries*. Journal of Materials Science: Materials in Electronics, 2005. 16: p. 799-807.

[36] Wang, Z. and Y. Xu, *Vibration energy harvesting device based on air-spaced piezoelectric cantilevers*. Applied Physics Letters, 2007. 90(26): p. 263512-263513.

[37] Kok, S.L., White, N. and Harris, N. Fabrication and characterisatin of free-standing, thick-film piezoelectric cantilevers for energy harvesting. Measurement Science and Technology, 20(124010).

[38] Larry, J.R., R.M. Rosenberg, and R.O. Uhler, *Thick-film technology: an introduction to the materials*. IEEE Trans.on Components, Hybrids, and Manufacturing Technology, 1980. CHMT-3(3): p. 211-225.

[39] Brignell, J.E., N.M. White, and A.W.J. Cranny, *Sensor applications of thick-film technology*. Communications, Speech and Vision, IEE Proceedings I, 1988. 135(4): p. 77-84.

[40] White, N.M. and J.E. Brignell, *A planar thick-film load cell*. Sensors and Actuators, 1991. 25 - 27: p. 313-319.

[41] Arshak, K.I., Ansari, F., McDonagh, D. and D. Collins, *Development of a novel thick-film strain gauge sensor system*. Measurement Science and Technology, 1997. 8(1): p. 58-70.

[42] Baudry, H., *Screen-printing piezoelectric devices*. Proc.6th European Microelec.Conf., 1987: p. 456-463.

[43] White, N.M., P. Glynne-Jones, and S.P. Beeby, *A novel thick-film piezoelectric micro-generator.* Smart Mater.Struct., 2001. 10: p. 850-852.

[44] Koplow, M., Chen, A., Steingart, D., P. Wright, and J. Evans. *Thick film thermoelectric energy harvesting systems for biomedical applications.* in *Proceedings of the 5th International Workshop on Wearable and Implantable Body Sensor Networks.* 2008.

[45] Hill, M., Townsend, R. J., Harris, N. R., White, N. M., S.P. Beeby, and J. Ding. *An ultrasonic MEMS particle separator with thick film piezoelectric actuation.* in *IEEE International Ultrasonics Symposium, 18-21 Sept 2005, Rotterdam, Netherlands.* Sept. 2005.

[46] Hale, J.M., White, J. R., R. Stephenson, and F. Liu, *Development of piezoelectric paint thick-film vibration sensors.* J. Mechanical Engineering Science, 2004. 219(1/2005).

[47] Schmid, S. and C. Hierold, *Two Sacrificial Layer Techniques for The Fabrication of Free-standing Polymer Micro Structures.* MicroMechanics Europe Workshop, Southampton, 2006: p. 177-180.

[48] Stecher, G., *Free Supporting Structures in Thick-Film Technology: A Substrate Integrated Pressure Sensor.* 6th European Microelectronics Conferences, Bournemouth, 1987: p. 421-427.

[49] Mitcheson, P.D., Yeatman, E. M., Rao, G. K., A.S. Holmes, and T.C. Green, *Energy Harvesting From Human and Machine Motion for Wireless Electronic Devices.* Proceedings of the IEEE, 2008. 96(9): p. 1457-1486.

[50] Glynne-Jones, P., El-Hami, M., Beeby, S., James, E. P., Brown, A. D., M. Hill, and N.M. White. *A vibration-powered generator for wireless microsystems.* in *Proc. Int. Symp. Smart Struct. Microsyst. Hong Kong.* Oct. 2000. Hong Kong.

[51] Roundy, S., P.K. Wright, and J.M. Rabaey, *Energy scavenging for wirless sensor networks.* Vol. 1st edition. 2003, Boston, MA: Kluwer Academic.

[52] Tanaka, H., Ono, G., T. Nagano, and H. Ohkubo, *Electric power generation using piezoelectric resonator for power-free sensor node,* in *Proc. IEEE Custom Integr. Circuits Conf.* 2005. p. 97 - 100.

[53] Reilly, K.E. and P. Wright. *Thin film piezoelectric energy scavenging systems for an on chip power supply.* in *Proc. Int. Workshop Micro Nanotechnlo. Power Generation Energy Conversion Applicat., Berkeley, CA.* Dec. 2006.

[54] Lefeuvre, E., Badel, A., Richard, C., Petit, L. and Guyomar, D., *A comparison between several vibration-powered piezoelectric generators for standalone systems.* Sensor and Actuators A, Phys., 2006. 126 (2): p. 405 - 416.

[55] Ferrari, M., Ferrari, V., D. Marioli, and A. Taroni, *Modeling, Fabrication and Performance Measurements of a Piezoelectric Energy Converter for Power Harvesting in Autonomous Microsystems.* IEEE Trans.on Instrumentation and Measurement, 2006. 55(6): p. 2096-2101.

[56] Mide. *Volture PEH25w online datasheet :* http://www.mide.com/pdfs/volture_specs_piezo_properties.pdf. [cited 2010 11 January].

[57] Ching, N.N.H., et al. *PCB integrated micro-generator for wireless.* in *Proc. Int. Symp. Smart Struct., Hong Kong SAR.* Oct. 2000.

[58] Li, W.J., Wen, Z., Wong, P. K., G.M.H. Chan, and P.H.W. Leong. *A micromachined vibration-induced power generator for low power sensors of robotic systems.* in *Proc. World Automat. Congr. 8th Int. Symp. Robot. Applicat., Maui, HI.* Jun. 2000.

[59] Williams, C.B., Shearwood, C., Harradine, M. A., Mellor, P. H., T.S. Birch, and R.B. Yates, *Development of an electromagnetic micro-generator*. IEE Proceedings - Circuits, Devices and Systems, 2001. 148(6): p. 337-342.

[60] Glynne-Jones, P., *Vibration powered generators for self-powered microsystems*, in *PhD Thesis, School of Electronics and Computer Science*. 2001, University of Southampton.

[61] Mizuno, M. and D.G. Chetwynd, *Investigation of a resonance microgenerator*. J. Micromech. Microeng., 2003. 13: p. 209 - 216.

[62] Huang, W.S., Tzeng, K. E., M.C. Cheng, and R.S. Huang, *A silicon MEMS micro power generator for wearable micro devices*. J. Chin. Inst. Eng., 2007. 30(1): p. 133 - 140.

[63] Torah, R., Glynne-Jones, P., Tudor, M., O'Donnell, T., S. Roy, and S. Beeby, *Self-powered autonomous wireless sensor node using vibration energy harvesting*. Measurement Science and Technology, 2008. 19(12): p. 125202.

[64] Ferro-Solutions. *VEH-360 online datasheet: http://www.ferrosi.com/files/VEH360_datasheet.pdf.* 2008 [cited 2010 12 Jan.].

[65] Perpetuum. *PMG37 online datasheet: http://www.perpetuum.com/resources/PMG37%20-%20Technical%20Datasheet.pdf.* 2009 [cited 2010 12 Jan].

[66] Tashiro, R., Kabei, N., Katayama, K., Ishizuka, Y.,, F. Tsuboi, and K. Tsuchiya, *Development of an electrostatic generator that harnesses the ventricular wall motion*. Jpn. Soc. Artif. Organs, 2002. 5: p. 239 - 245.

[67] Arakawa, Y., Y. Suzuki, and N. Kasagi. *Micro seismic power generator using electret polymer film*. in *Proc. 4th Int. Workship Micro and Nanotechnology for Power Generation and Energy Conversion Applicat. Power MEMS*, Kyoto, Japan. Nov. 2004.

[68] Despesse, G., Chaillout, J., Jager, T., Leger, J. M., Vassilev, A., S. Basrour, and B. Charlot. *High damping electrostatic system for vibraiton energy scavenging*. in *Proc. 2005 Joint Conf. Smart Objects Ambient Intell. -Innov. Context-Aware Services: Usages Technolo.*, Grenoble, France. 2005.

[69] Miao, P., Mitcheson, P. D., Holmes, A. S., Yeatman, E. M., T.C. Green, and B. Stark, *MEMS inertial power generators for biomedical applications*. Microsyst Technol, 2006. 12(10-11): p. 1079-1083.

[70] Basset, P., Galayko, D., Mahmood Paracha, A., Marty, F., A. Dudka, and T. Bourouina, *A batch-fabricated and electret-free silicon electrostatic vibration energy harvester*. J. Micromech. Microeng., 2009. 19.

6

Wearable Energy Harvesting System for Powering Wireless Devices

Yen Kheng Tan and Wee Song Koh
Energy Research Institute @ NTU (ERI@N)
Singapore

1. Introduction

As the world trends towards ageing population UN (2011), there is an increasing demand and interest in using technology to increase the quality of life for elderly people. An expanding area of interest is heading towards the health care applications like wearable biometric monitoring sensors. These monitoring nodes, typically powered by batteries, have various functions like sensing & monitoring bodily functions, after which the data is wirelessly transmitted to a remote data terminal Harry et al. (2009), Philippe et al. (2009). However such applications mentioned are not new, where earlier literatures envisioned of a not too distant future where e-textiles, electronics woven together with fabrics, are omni-present Marculescu et al. (2003). With improving technology in miniaturization and wireless communication, clothing containing sensors for sensing and monitoring bodily physiological functions Wixted et al. (2007) is becoming more common and widespread. Such devices should be unobtrusive wearable, flexible, lightweight and ideally self-sufficient.

In using batteries, the useful life of a wearable sensing device Cook et al. (2004) is usually limited by the battery's lifespan or capacity. Using a high energy capacity AA sized battery of 3000mAh, the life of battery powering a certain sensor node can last a maximum of 1.5 years Kheng et al. (2010). But operation life of the wearable electronic is much longer, at least several years. Therefore its normal operation will be interrupted whenever the supplying batteries die out. Typically, the higher the capacity of the battery, bigger in size the battery will be. With miniaturization, device components like sensors, accompanying electronics and board size will shrink and get smaller. As such, wearable flexible batteries are more commonly used to replace the larger batteries to keep pace with the shrinkage of these wearable electronics. But capacity of a flexible thin-film battery with a volumetric size of 1.2 cm^3 is about 30 mAh, lower than a 2850mAh capacity AA alkaline battery of volumetric size 11 cm^3. As a result, sustainability is often a key challenge for systems to be standalone with 'Deploy & Forget' feature.

The addition of energy harvesting source is identified as a feasible way to increase the device's operation duration. Several potential ambient energy sources are discussed, with the photovoltaic (PV) harvesting method providing the highest power density per volume of total system Raghunathan et al. (2005). For indoor application using PV harvesting, major challenges include: poor lighting intensity as compared to outdoor lighting intensity; limited sized PV panel to be used if the device is to be placed in a confined area of the human body. Such power produced by the PV panel is very small, usually in the range of hundreds of

μW. Indoor light intensity in office environment is in the range of < 10 W/m^2, compare with 100-1000 W/m^2 for outdoor conditions Hande et al. (2007). For photovoltaic panel, amorphous type is the best suited for indoor applications but suffers from low efficiency, in the range of 3% - 7% Randall et al. (2002). On the other hand, PV cells provide a fairly stable DC voltage through much of their operating space Roundy et al. (2004). Various works utilizing solar harvesting/scavenging techniques demonstrated the suitable of indoor PV in supplying alternate energy to small, low-power consuming devices, complemented with batteries within the system Hande et al. (2007), Nasiri et al. (2009).

In this chapter, a flexible and self-sustainable energy solution incorporating energy harvesting for wearable electronics is presented. Introducing of energy harvesting technique levitate the operation of system towards self-sustainability. However for the system to be wearable, certain amount of device flexibility or bending is needed. Generally the device should not be rigid, not inhibit motion in any way and ideally follow as closely to the contour of the wearer's body. As such, in this chapter, rigid batteries like the AA size battery, PV panel, PCB and supercapacitor are replaced with the flexible, bendable version. Capacity of a typical flexible battery is in the range of a few tens of mAh Hahn et al. (1999), which severely restrict the node's operation duration if the flexible battery is the only input source. As such, an additional input source is hybrid with the primary battery (which in this case, PV panel is chosen as the additional input source to complement the primary battery for powering the wireless body sensors). Flexible super capacitors with capacitance of ≈ 11 F/g has been realized with good capacitance stability for long term usage applications Gan et al. (2009). The rest of the chapter is organized as follows: Section II introduces the wearable energy storage for wireless body sensor network and section III illustrates in more details about the key part of the proposed system: flexible energy harvesting system comprising of modules like maximum power point tracking (MPPT), current limiter, voltage regulation within the power management circuitry and the load requirements. After which, in section IV, the hybrid of wearable energy storage and FEH is discussed. Experimental results of the proposed system performance are illustrated in section V and conclude the chapter with section VI.

2. Wearable energy storage for wireless body sensor network

It is anticipated that people will soon be able to carry a personal body sensor network (WBSN) system with them that will provide users with information and various reporting capabilities for medical, lifestyle, assisted living, sports or entertainment purposes. In the literature, some older medical monitoring systems (such as Holter monitors) record the hosts' data for off-line processing and analysis. Newer wearable wireless systems provide almost instantaneous information that help in earlier detection of abnormal conditions. There are also many such commercial products out there to allow wearers to monitor their vital signs, for examples, Omron health care products like blood pressure meter, thermometer and portable ECG and Philips vital sense product and sports monitoring devices as seen in Figures.1 and 2. For these commercially available health care products as seen in Figure.1, although they are meant to be made for small size and portable, in actual fact, they are too big and bulky to be integrated as part of our bodies for monitoring. Part of the reason why these products are so huge is because of the batteries. Moreover, these products operate heavily on their onboard batteries and if they are to conduct continuous body monitoring, their operational lifetimes are very short, a month or even less than that.

Fig. 1. Omron healthcare products (a) portable ECG, (b) thermometer and (c) blood pressure meter and Philips product (d) vital sense device

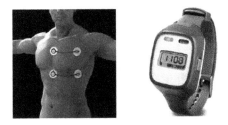

Fig. 2. Body worn devices for measuring activity and energy expenditure

Having said that, these body worn devices are still receiving huge attentions and commercial demands simply because of their outstanding features, but they really need to be highly portable and easily embeddable into our bodies for monitoring. In addition to that, the catch with these body worn devices is the sky high prices to own an outstanding system like this, i.e. a few hundreds or even to a thousand dollars. If there are a few more places on the human body for close measuring and monitoring of dedicated activities like sleeping, sporting, etc., it will cost a huge sum to implement the body monitoring system. There is no doubt about the potential of such body worn monitoring system and the market is huge demanding for such distributed sensing of human well beings through their vital signs. However, the present state of arts and commercial products are limited and there are more to what they have that could be included. As compared to the conventional large and bulky body monitoring system mentioned earlier, the availability of microelectronics devices and micro electromechanical systems (MEMS) like pulse oximeters, accelerometers, energy harvesting devices, etc. Wixted et al. (2007), Cook et al. (2004) integrated with wireless technology provides an alternative, non-invasive, distributed and self-powered method of automatic monitoring activity. In addition, many of such miniaturized electronic devices are integrated together into each individual person and also into their activities to enable better human-computer interaction to achieve all-rounded monitoring of human health lifestyle and more accurate performance assessment of the athletes as illustrated in Figure.3.

The functionality of the proposed body monitoring system in Figure.3 on each individual human being is illustrated as follows: the sensed physiological information of the human is stored and accumulated in the memory of the sub-GHz ultra-wide-band (UWB) transmitter and it is periodically communicated to the UWB receiver of the base station without mutual interference. One of the approaches is by coding the sequence or using different time slots, the receiver can identify the transmitter from which sensor and setup the link automatically. The received data from various smart sensors deployed around the body are then used for performance assessment of subject under test. Wireless communication does away the wires, hence save the wearer of this proposed body monitoring system from the phobia of wires.

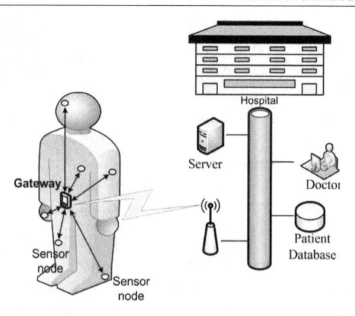

Fig. 3. Human health lifestyle monitoring

Even though wires are removed, battery becomes the concern as the operational lifetime of the energy storage is limited. The effective duration of a battery driven body monitoring system is short in terms of days of weeks, after which the monitoring purpose is gone.

The energy problem escalates further when there is a need for the energy storage to be flexible and wearable, able to conform to human body. According to the authors of Harry et al. (2009) and Philippe et al. (2009), both suggested the use of thin-film battery technology to shrink the overall package size, where lithium polymer battery sizes of 85 mm x 55 mm x 0.5 mm and 59 mm x 35 mm x 0.5 mm (PGEB0053559) to achieve the wearable energy storages. Typical flexible (thin film solid state) batteries are constructed by depositing the components of the battery as thin films (usually in tens of μm) on a substrate, which includes a solid substrate of electrolyte cathode (positive electrode) and anode (negative electrode). Advantages include small physical size, able to be used in a very broad range of temperatures, and supposedly more eco-friendly than conventional batteries Mcdonald (2011). However, as with all batteries applied on WBSN, they will be drained off after a certain period of time. In Harry et al. (2009) and Philippe et al. (2009), rechargeable lithium polymer battery capacity is of 50 to 200 mAh (12 hours to 50 hours of operation) and 65 mAh at 3.7 V respectively. Clearly, wearable energy storage alone is not able to sustain the operation of the WBSN. There is a need to seek for a supplement flexible energy harvesting system to prolong the operational lifetime of the WBSN.

3. Flexbile energy harvesting system

To minimize the problem associated with batteries, using of photovoltaic as an addition energy source is proposed as a solution to complement battery (Zn-MnO$_2$ flexible battery

Barbic et al. (1999), rated voltage at about 1.5 V and capacity of $\approx$ 30 mAh) in prototype and to prolong the operational life of the wearable device.

3.1 Characteristics of PV panel

Photovoltaic cell converts light to electricity through a physical process called the photovoltaic effect. Light (in the form of photons) that is absorbed into the PV cell will transfer its energy to the semiconductor device, knocking electrons loose and allowing them to flow freely. These generated electrons are transferred between different bands (example, from the valence to conduction bands) within the material, resulting in the buildup of voltage between two electrodes. Electrically, a solar cell is equivalent to a current generator in parallel with an asymmetric, non-linear resistive element (example: a diode). When illuminated, the ideal cell will produce a photocurrent proportional to the light intensity. That photocurrent is divided between the variable resistance of the diode and the load, in a ratio which depends on the resistance of the load and the level of illumination. For higher resistances, more of the photocurrent flows through the diode, resulting in a higher potential difference between the cell terminals but a smaller current though the load. The diode thus provides the photovoltage. Without the diode, there is nothing to drive the photocurrent through the load Nelson (2011).

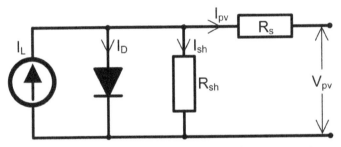

Fig. 4. Equivalent electrical circuit for a photovoltaic cell with parasite resistances

Figure.4 shows the basic equivalent circuit of a PV cell, where I_L - light-generated current, I_D - reverse saturation (dark) current of the PN diode, R_s - series resistance, R_{sh} - shunt resistance. Dark current can be viewed as caused by the potential built up over the load and flows in the opposite direction. When the shunt resistance, R_{sh} is assumed to be infinite, the current-voltage (I-V) characteristic of the photovoltaic (PV) module can be described with a single diode as the four-parameter model given by,

$$I_{pv} = I_L - I_D \left[exp \left(\frac{V_{pv} + I_{pv} R_s}{N_s n_I V_t} \right) - 1 \right] \tag{1}$$

where V_t - the junction terminal voltage, N_s is the number of cells in series and n_I is the diode ideality factor Celik (2007). For this prototype, off-the-shelf Sundance Solar MPT3.6-75 Sundance (2011) flexible PV panels, made up of amorphous silicon on a polymer substrate, is used. Dimensions are about 75 mm x 72 mm x 0.5 mm. PV characterization graphs are shown in Figures.5 and 6. At $\approx$ 400Lux, it is able to provide a peak power of about 0.14 mW.

Any unused energy will be stored into a flexible supercapacitor, which is ideal for energy storage that undergoes frequent charge and discharge cycles at high current and short

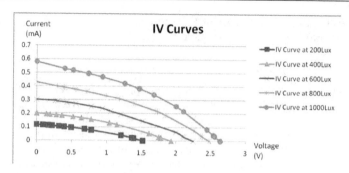

Fig. 5. IV Curves of PV panel at various Lux values

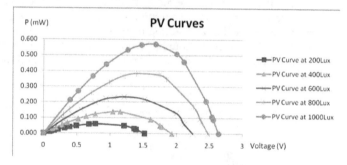

Fig. 6. PV Curves of PV panel at various Lux values

Fig. 7. A flexible supercapacitor laminated using polymer-coated aluminum foil

duration. Basically the plates of a supercapacitor are filled with two layers of the identical substance for separating the charge, instead of having dielectric, resulting in a much larger surface area and high capacitance. Experiments using various types of electrodes and electrolyte had been extensively carried out, like experimenting VNF electrodes in aqueous electrolyte of different pH and also in an organic electrolyte Grace et al. (2010). Dimension of such flexible capacitors as shown in Figure.7 can be packaged to about the same size as the flexible battery.

3.2 Fractional open-circuit voltage MPPT technique

Maximum Power Point Tracking (MPPT) is a frequently used technique to vary the electrical operating point of the PV module so that the module is able to deliver its maximum available power. Various MPPT techniques are grouped into 'Direct' or 'Indirect' methods Salas et al. (2005). For indirect methods ("quasi seeks"), the Maximum Power Point (MPP) is estimated from the measures of the PV generator's voltage and current PV, the irradiance, or using empiric data, by mathematical expressions of numerical approximations. They do not obtain the maximum power for any irradiance or temperature and none of them are able to obtain the MPP exactly. But in many cases, such methods can be simple and inexpensive. The direct methods ("true seeking methods") obtain the actual maximum power from the measures of the PV generator's voltage and current PV. Although Fractional Open Circuit Voltage based MPPT method is classified as a quasi seeks method, it is also considered to be one of the simplest and cost effective method Masoum et al. (1999). It is based on the fact that the PV array voltage corresponding to the maximum power exhibits a linear dependence with respect to the array open circuit voltage for different irradiation and temperature levels. Maximum power point voltage, $V_{MPP} = K_{oc} * V_{oc}$, where V_{oc} is the open circuit voltage of the PV and K_{oc} is the voltage factor Ahmad (2010). To operate the PV panel at the MPP, the actual PV array voltage V_{pv} is compared with the reference voltage V_{ref} which corresponds to the V_{mpp}. The error signal is then processed to make $V_{pv} = V_{ref}$. Normally, the panel is disconnected from the load momentarily to sample its open circuit voltage. The fraction of the open circuit voltage corresponding to the V_{mpp} is measured and is kept in a hold circuit to function as V_{ref} for the control loop.

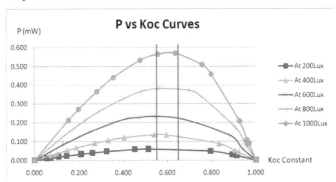

Fig. 8. Graph of Power vs Koc Constants

In Figure 8, the peak power of the PV panel is found between K_{oc} constant values of 0.55 to 0.65. The K_{oc} part of the control circuit will reference a K_{oc} constant of 0.65 to V_{oc} as V_{ref}. The control circuit will be built using discrete components and op-amps.

3.3 Ultra-low-power management circuit

The MPPT control circuitry block diagram is shown in Figure.9. It is designed to boost V_{pv} to the load when it has fallen below the V_{ref} reference value. First the PV panel will break open from rest of circuit by means of a switch. This open circuit voltage will be captured by the K_{oc} circuit, multiplied by the K_{oc} constant to become V_{ref}. After a certain time interval, the PV panel will connect back with the rest of the circuit. If $V_{pv} < V_{ref}$, the error signal will be amplified and compared with a sawtooth waveform, with the resultant signal controlling the gate of the DC-DC converter.

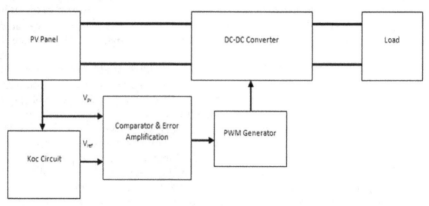

Fig. 9. Block diagram of the Fractional Open Circuit Voltage MPPT control circuit

Using discrete components to build this control circuit, MOSFET are used as switches where timing switching will be controlled by pre-programmed pulses from MSP430 MCU onboard the end device. Koc constant of 0.65 is obtained using voltage dividing in the Koc circuit. Op-amps, capacitors, resistors and Schottky diodes are used in various part of the circuit for comparisons and simple sample & hold operations.

3.4 Wireless body sensor nodes/network

The wireless body sensor node is developed from the target board from Texas Instruments eZ430-RF2500 Development Tool Texas Instruments (2011), which measures the body temperature of the wearer and communicates wirelessly to an access point connected to a PC. It operates between 1.8 V to 3.6 V, and measured 35 mm x 20 mm x 3.5 mm, which can be easily placed into cloths pocket or between layers of sewn clothing. The communication profile is captured in Figure.10.

Referring to Figure.10, during the sleep/standby mode, the target board consumes around 1.2 μA of quiescent current. During initialize stage, instantaneous current can rise up to about 20 mA and 2 mA for burst mode transmission, which is taken care of by the flexible supercapacitor. To reduce current consumption by the load, the target board had been configured to transmit data at a $\approx$ 5 seconds transmission period. In its original mode, its average current consumption over 1 second transmission period is 36.80 μA Texas Instruments (2011).

$$I_{ave} = [I_{sleep} + I_{Tx,Total}]/T_{Tx} \qquad (2)$$

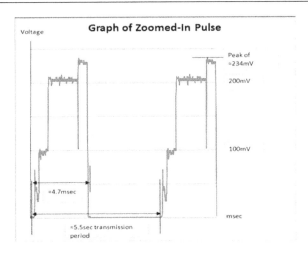

Fig. 10. Zoomed-In pulse current consumption by the TI Target Board with modified configuration of ≈ 5 second transmission period. Graph captured across a 10Ω current sensing resistor in series with load

where

$$I_{sleep} = [I_{idle}(MSP430) + I_{idle}(CC2500)] * [T_{Tx}(sec) - T_{app}(sec)] \qquad (3)$$
$$= 1.3[\mu A] * (1[s] - 2.838[ms])$$
$$= 1.296A * s$$

Therefore average current consumption over 1 second transmission period: I_{ave} (1 sec Tx period) = (1.296 [μA*s] + 35.508 [μA*s]) / 1 [s] = 36.80 μA. If transmission period is extended to ~5 second, the sleep current: I_{sleep} (over 5 sec Tx period) = 1.3 [ţA] * (5 [s] - 2.838 [ms]) = 6.496 μA*s and the average current consumption over 1 second: I_{ave} (over 1 sec for a 5sec Tx period) = (6.496 [μA*s] + 35.508 [μA*s]) / 5 [s] = 8.4 μA

Therefore the less frequently the target board transmits, the less average current is consumed.

If using battery of 1000 mAh capacity for average current of 36.80 μA consumption, calculated life expectancy = 1000 [mA*hrs] / 0.0368 [mA] ≈ 3.10 years. If using battery of 30 mAh capacity for average current of 36.80 μA consumption, calculated life expectancy = 30 [mA*hrs] / 0.03680 [mA] ≈ 33.97 days. If using battery of 30mAh capacity for average current of 8.4 μA consumption, calculated life expectancy = 30 [mA*hrs] / 0.0084 [mA] ≈ 148.8 days.

Such calculations constitute an perfect scenario, where there is no leakages, 100 % efficiency, no power loss, no surges, and can only be used as a rough guide in the calculation of the life expectancy of battery of certain capacity, and the actual current consumption of the node. In the worst case scenario, the end device will keep initializing when scanning to linkup with the access point, which will continuously draw about 18 mA of current.

4. Hybrid flexible energy harvesting and energy storage

The proposed hybrid flexible energy system prototype as seen in Figure.11 incorporate three different types of energy sources, mainly the primary battery (flexible batteries), the secondary battery (flexible supercapacitor, which acts as energy storage) as well as renewal energy harvesting source like the flexible PV panel (Additional input energy to complement primary batteries) to harvest ambient light energy.

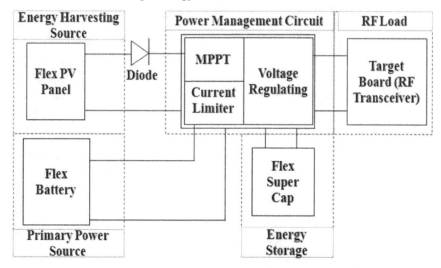

Fig. 11. Block diagram of the proposed hybrid energy harvesting and storage system

The RF transceiver load typically has 2 modes of operation: sleep and transmission. In sleep mode, it consumes around 1.2 μA of quiescent current, and in transmission mode, it consumes around 2 mA of current. The default transceiver setting is set at 1 transmission for every 1 second period. During sleep mode, the transceiver consumes very little energy, where excess unused energy from the primary battery and PV panel will be stored in the flexible supercapacitor. During transmission mode, the transceiver will draw energy mainly from the supercapacitor, which is able to manage the sudden current surge. Subsequently, the transceiver goes into sleep mode and the supercapacitor starts to recharge from the primary battery and PV panel. To conserve energy, period between transmissions is increasing, which decrease energy usage and increase the charging time for supercapacitor.

The overall system design of the proposed hybrid flexible energy harvesting and storage solution is illustrated in Figure.11. The power management circuitry of the system depicted in Figure.11 consists of: MPPT control, flexible battery current limiter and a load voltage regulator, to be fabricated onto a flexible PCB substrate. MPPT control provides a simple mean of impedance matching between PV panels to load. The current limiter protects the primary battery from sudden surges. Voltage regulator maintains a steady voltage level as required by the transceiver load. Referring to Figure.12, the MPPT control circuit is designed to boost V_{pv} to the load when it has fallen below the V_{ref} value. To capture the open circuit voltage of the PV panel, NMOS 1 will be opened to isolate the PV panel from rest of the circuit,

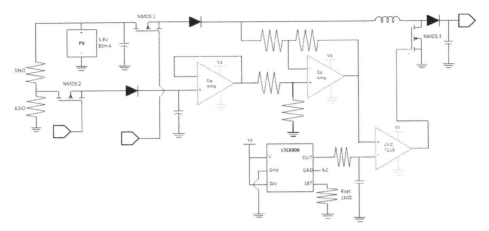

Fig. 12. MPPT Control Circuit using Fractional Open Circuit Voltage Approach

while NMOS 2 will be closed. The open circuit voltage will be acquired by the K_{oc} circuit, which is a voltage dividing circuit, and multiplied by the predetermined K_{oc} constant of 0.65 to become V_{ref}. Subsequently this V_{ref} value is stored by an op amp sample and hold circuit. Pre-programmed timers signal from MSP430 MCU onboard the TI end device will provide the switching timing for NMOS 1 and NMOS 2. After a set timing, NMOS 2 will be opened whereas NMOS 1 will be closed. The voltage from the PV panel, V_{pv}, will be compared with the V_{ref} at the second op amp, which is a differential amplifier. When V_{pv} is lower than V_{ref}, a voltage output will be sent to the '+' input of the LMC7215 comparator, which will compare with the sawtooth waveform from LTC6906. This will provide switching for NMOS 3 of the boost converter.

5. Experimental results

5.1 Performance of flexible energy harvesting system with MPPT scheme

In switching the gate of NMOS 3, there must be output from LMC7215, meaning the sample & hold voltage (V_{ref}) must be more than voltage of PV panel (V_{pv}). For example, shading over the PV panel occurs, causing V_{ref} to be more than V_{pv}, and switching at NMOS 3 gate to commence. Typically the greater the difference between V_{ref} and V_{pv}, the larger the duty cycle of the switching signal will be.

At $\approx$ 320 lux, the PV panel shows an open circuit voltage of about 1.36 V and V_{pv} voltage of about 0.3 V (Figure.13). At this lux level, the maximum power that the PV panel is able to produce is about 76 μW, which correspond to around 0.8 V and 0.1 mA on the PV and IV graphs. This maximum power point voltage is captured by the sample and hold circuit. When connected to rest of the circuit, the V_{pv} voltage drops to about 0.3 V, which corresponds to $\approx$ 40 μW. There is a further voltage drop of $\approx$ 0.11 V drop across the Schottky diode. Therefore the input voltage to the boost converter is around 200 mV.

The configuration of the differential op amp will influence the duty cycle to the gate of NMOS 3, which will in turn determine the output voltage of the boost converter. In an earlier configuration, the op amp has been configured to give an output voltage where $V_{out} = 1/3$

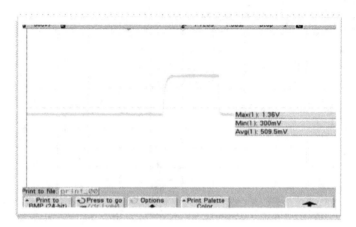

Fig. 13. Voltage across PV panel

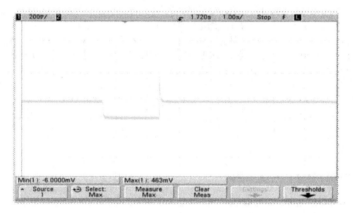

Fig. 14. Voltage waveform at input of boost converter

[V(+) - V(-)]. V(+) is about 0.8 V, while V(-) is about 0.3 V. V_{out} of the differential op amp is about 0.167 V. When compared with the sawtooth waveform, duty cycle of around 30% is produced to NMOS 3. Using boost converter formulae, $V_{out} = V_{in}/(1 − D)$, output voltage of converter is about 300 mV (200 mV / 0.7) when its input voltage is around 200 mV, as shown in Figures.14-16.

5.2 Power conversion efficiency of FEH system

At 320 lux, the PV panel produces $P_{pv} = V_{pv} * I_{pv} = 300mV * 0.11mA = 0.033mW$ when connected to rest of circuitry. At input of boost converter, P_{in} = V*I = 190 mV*0.1 mA = 19 μW. Difference in power between P_{pv} and Power at Boost converter input is due to the voltage drop across the Schottky diode after NMOS 1. At output, $P_{out} = V_{load} * I_o$ = 300 mV*36 μA = 10.8 μW. Efficiency of the boost converter, η = 10.8 μW/19 μW ≈ 60%.

At 320 lux, G = 320/120 ≈ 2.67 W/m^2.

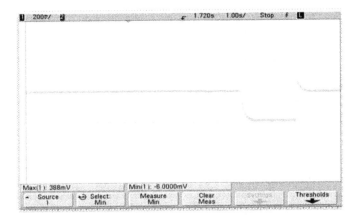

Fig. 15. Waveform at output load

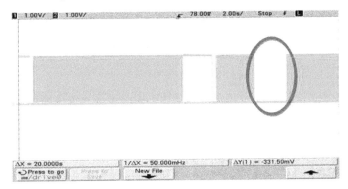

Fig. 16. No switching at Gate of NMOS3 when V_{ref} is less than V_{pv}

Parameter	Unit	Value
Cross-section area	cm²	7.2 x 5.9 = 42.48
OC voltage	V	1.46
SC current	μA	155
MPPT voltage	V	0.886
MPPT current	μA	86.3

Table 1. Technical Characteristic of PV Panel used

Efficiency of flexible PV panel:

$$\eta_{pv} = \frac{P_{pv}}{G * A} * 100\% \qquad (4)$$
$$= [(0.886V * 86.3\mu A)/(2.67W/m^2 * 42.48cm^2)] * 100\%$$
$$= 0.73\%$$

At 400 lux, the maximum power that the PV panel is able to produce is about 130 μW, which correspond to around 1 V and 0.13 mA on the PV and IV graphs. At output of boost converter, current is about 40 μA. Over a 5 second period, total current accumulated is 200 μA.sec, which is able to fulfill the 0.2 mA needed by the end device for transmission in a 5 second period.

5.2.1 Power consumption study of discrete components within the fractional open circuit voltage approach control circuit

When using a 555-timer and op-amp inverter (MAX9077) to control the opening and closing for NMOS 1 and 2, the power consumption is shown in the next table.

	mA		Voltage (V), from Power Supply	Power (mW)	% of current consumed by
Total system current consumption	0.27		2.5	0.675	
CMOS 555 TIMER	0.121		2.5	0.3025	44.81%
MAX9077	0.0053	5.3uA	2.5	0.01325	1.96%
LTC6906	0.00984	9.84uA	2.5	0.0246	3.64%
LMC7215	0.0065	6.5uA	2.5	0.01625	2.41%
MAX4471 (2 Op Amps)	0.04	40uA	2.5	0.1	14.81%

Table 2. MPPT control circuit power consumption at 2.5V

From Table.2, both components (555-timer and MAX9077) take up $\approx$ 40 % of the total MPPT control circuit power consumption. We can decrease the total discrete components power consumption within the MPPT control circuit by using the RF transceiver's MCU to provide the timing pulse function to control NMOS 1 and 2.

5.2.2 Current limiter

A current limiter, LM334 National Semiconductor (2011) is added to protect the flexible battery from any massive drain due to possible surge in load. At 0.14mA limit, the flexible is able to take care of the quiescent current of the target board. If constant drawing of 0.14 mA from a fully charged 30 mAh flexible primary battery, calculated life expectancy of flexible battery will be: Duration = 30 [mA*hrs] / 0.14 [mA] $\approx$ 8.93 days

5.2.3 Voltage regulation

The voltage regulator, LTC3525-3.3 Linear Tech. (2011) is a compact, step-up DC-DC converters used to regulate the output voltage to the target board. The regulator has a V_{in} range of 0.5 V to 4.5 V, fixed output voltage of 3.3V and capable of delivering 60 mA at 3.3 V from a 1 V input. Quiescent current is an ultra-low 7 μA, maximizing battery life in portable applications. Its efficiency verse load current drawn is shown in Figure 17.

5.3 Application of hybrid FEH and storage for wearable wireless body sensor network

In the placement of the prototype, the flexible batteries and supercapacitor are flexible enough to bend along the contours of the human body, like on the forearm and shoulder.

Unlike flexible supercapacitor which has higher capacitance when twisted than any non-twisted supercapacitor Zyga (2011), flexible PV panel will see a further decrease in efficiency if bended as solar irradiance will decrease with less PV panel surface exposed directly to the light source (see Figure.20).

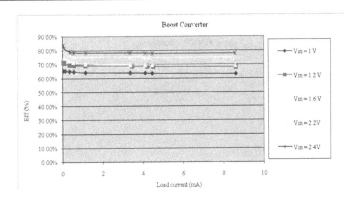

Fig. 17. Efficiency verse load current graph of boost converter

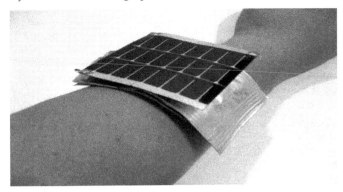

Fig. 18. Prototype placed and wrapped around the forearm

Fig. 19. Prototype placed at shoulder

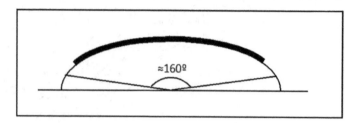

Fig. 20. Flexible PV panel placed on an arc with circular angle of 160ž

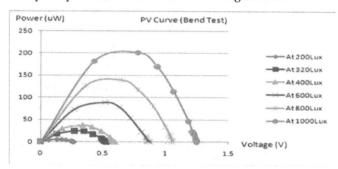

Fig. 21. PV Curves of PV panel at various Lux values when subjected to bending

From Figure.21, power produced dropped to about 1/3 of that from a flat panel. However as a starting platform, functional experiment is conducted with a flat prototype in a controlled environment.

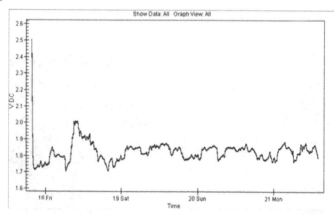

Fig. 22. Constant super capacitor voltage of around 1.7 V under a constant light source of 400 lux

Referring to Figure.22, it can be seen that the functionality experiment of prototype, under a constant light source of ≈ 400 lux, the system is self sustainable, powering the target board and maintaining a super capacitor voltage of around 1.7 V. The experimental result verifies that the proposed hybrid flexible energy harvesting and storage system is able to sustain the operation of a wireless body sensor node connected in a network form.

6. Conclusions

Utilization of the PV harvesting prototypes together with its individual parts has been introduced. Under constant light intensity of 400 lux in a controlled environment, the prototype is able to operate in a self-sustaining mode, where the super capacitor voltage maintains at around 1.7 V. With its flexible attribute, it can be expanded to wearable, biomedical, constraint space applications and subsequent flexible design can be modified to incorporate other harvesting techniques.

7. References

United Nations, Department of Economic and Social Affairs, Population Division, "World Population Ageing: 1950-2050", >http://www.un.org/esa /population /publications/worldageing19502050/< assessed on 02-09-2011.

Harry K. Charles, Jr. and Russell P. Cain, "Ultra-thin, and Flexible Physiological Monitoring System", *IEEE Sensors Applications Symposium,* 2009

Philippe Jourand, Hans De Clercq, Rogier Corthout, Robert Puers, "Textile Integrated Breathing and ECG Monitoring System", *Proceedings of the Eurosensors XXIII conference, Procedia Chemistry 1*, pp.722-725, 2009.

D. Marculescu, R. Marculescu, S. Park and S. Jayaraman, "Ready to Ware", *IEEE Spectrum*, pp.28-32, 2003.

A.J. Wixted, D.V. Thiel, A.G. Hahn, C.J. Gore, D.B. Pyne and D.A. James, "Measurement of Energy Expenditure in Elite Athletes Using MEMS-Based Triaxial Accelerometers", *IEEE Sensors Journal*, vol.7, no.4, pp.481-488, 2007.

D.J. Cook and S. K. Das, "Wireless Sensor Networks - Smart Environments: Technologies, Protocols and Applications", *John Wiley*, New York, 2004.

Yen Kheng Tan, Sanjib Kumar Panda, "Review of Energy Harvesting Technologies for Sustainable Wireless Sensor Network", Sustainable Wireless Sensor Networks, *INTECH Publisher*, Chap 2, pp.15-43, 2010.

V. Raghunathan, A. Kansal, J. Hsu, J. Friedman and M. Srivastava, "Design Considerations for Solar Energy Harvesting Wireless Embedded Systems", *Fourth International Symposium on Information Processing in Sensor Networks Proceedings*, pp.457-462, 2005.

A. Hande, T. Polk, W. Walker, and D. Bhatia, "Indoor Solar Energy Harvesting for Sensor Network Router Nodes", Microprocess. Microsyst., vol.31, no.6, pp.420-432, 2007.

J.F. Randall, J. Jacot, "The Performance and Modelling of 8 Photovoltaic Materials under Variable Light Intensity and Spectra", *World Renewable Energy Conference VII Proceedings*, Cologne, Germany, 2002.

S. Roundy, P.K. Wright, J.M. Rabaey, "Energy Scavenging for Wireless Sensor Networks: with Special Focus on Vibrations". *Kluwer Academic Publishers*, 2004.

Adel Nasiri, Salaheddin A. Zabalawi, Goran Mandic, "Indoor Power Harvesting Using Photovoltaic Cells for Low-Power Applications", *IEEE Transactions on Industrial Electronics*, vol.56, no.11, 2009.

Robert Hahn, Herbert Reichl, "Batteries and Power Supplies for Wearable and Ubiquitous Computing", *Third International Symposium on Wearable Computers*, Digest of Papers, 1999.

Hiong Yap Gan, Cheng Hwee Chua, Soon Mei Chan and Boon Keng Lok, "Performance Characterization of Flexible Printed Supercapacitors", *11th Electronics Packaging Technology Conference*, 2009.

Jason Mcdonald, "Thin film battery technology and advanced batteries", >http://www.eg3.com/blog/20090420.htm< assessed on 02-09-2011.

P.A. Barbic, L. Binder, S. Voss, F. Hofer and W. Grogger, "Thin-Film Zinc/Manganese Dioxide Electrodes based on Microporous Polymer Foils", *Journal of Power Sources*, vol.79, issue.2, pp.271-276, 1999.

Jenny Nelson, "The Physics of Solar Cells", *Imperial College Press*, >http://www.worldscibooks.com/ physics/p276.htm< assessed on 02-09-2011.

Ali Naci Celik, Nasir Acikgoz, "Modelling and Experimental Verification of the operating current of mono-crystalline photovoltaic modules using four- and five-parameter models", *Applied Energy*, vol.84, issue.1, pp.1-15, 2007.

Sundance Solar, "Small Solar Panels for science fair projects, experiments and prototypes", >http://store.sundancesolar.com/smalsolpanfo.html< assessed on 02-09-2011.

Grace Wee, Oscar Larsson, Madhavi Srinivasan, Magnus Berggren, Xavier Crispin, Subodh Mhaisalkar, "Effect of the Ionic Conductivity on the Performance of Polyelectrolyte-Based Supercapacitors", *Advanced Functional Materials*, pp.4344-4350, 2010.

V. Salas, E. Olit'as, A. Barrado, A. Lat'zaro, "Review of the Maximum Power Point Tracking Algorithms for Stand-alone Photovoltaic Systems", *Solar Energy Materials & Solar Cells* 90 (2006), pp.1555-1578, 2005.

M.A.S. Masoum, H. Dehbonei, "Design Construction and Testing of a Voltage-based Maximum Power Point Tracker for Small Satellite Power Supply", *Proceedings of 13th annual AIAA/USU Conference on Small Satellite*, 1999.

Jawad Ahmad, "A Fractional Open Circuit Voltage Based Maximum Power Point Tracker for Photovoltaic Arrays", *2nd International Conference on Software Technology and Engineering (ICSTE)*, 2010.

Texas Instruments, eZ430-RF2500 Development Tool, Datasheet.

Texas Instruments, eZ430-RF2500 Development Tool, Application Notes.

National Semiconductor, LM334, Datasheet.

Linear Technology, LTC3525-3.3, Datasheet.

Lisa Zyga, "Paper-thin supercapacitor has higher capacitance when twisted than any non-twisted supercapacitor", >http://www.physorg.com/news204265367.html< assessed on 02-09-2011.

Q. Zhang, P. Feng, Z.Q. Geng, X.Z. Yan, N.J. Wu, "A 2.4-GHz energy-efficient transmitter for wireless medical applications", *IEEE Transactions on Biomedical Circuits and Systems*, vol.5, no.1, pp.39-47, 2011.

O. Omeni, Alan C.W. Wong, Alison J. Burdett, C. Toumazou, "Energy efficient medium access protocol for wireless medical body area sensor networks", *IEEE Transactions on Biomedical Circuits and Systems*, vol.2, no.4, pp.251-259, 2008.

7

WSN Design for Unlimited Lifetime

Emanuele Lattanzi and Alessandro Bogliolo
DiSBeF - University of Urbino
Italy

1. Introduction

Wireless sensor networks (WSNs) are among the most natural applications of energy harvesting techniques. Sensor nodes, in fact, are usually deployed in harsh environments with no infrastructured power supply and they are often scattered over wide areas where human intervention is difficult and expensive, if not impossible at all. As a consequence, their actual lifetime is limited by the duration of their batteries, so that most of the research efforts in the field of WSNs have been devoted so far to lifetime maximization by means of the joint application of low-power design, dynamic power management, and energy-aware routing algorithms. The capability of harvesting renewable power from the environment provides the opportunity of granting unbounded lifetime to sensor nodes, thus overcoming the limitations of battery-operated WSNs. In order to optimally exploit the potential of energy-harvesting WSNs (hereafter denoted by EH-WSNs) a paradigm shift is required from energy-constrained lifetime maximization (typical of battery-operated systems) to power-constrained workload maximization. As long as the average workload at each node can be sustained by the average power it takes from the environment (and environmental power variations are suitably filtered-out by its onboard energy buffer) the node can keep working for an unlimited amount of time. Hence, the main design goal for an EH-WSN becomes the maximization of its sustainable workload, which is strongly affected by the routing algorithm adopted.

It has been shown that EH-WSNs can be modeled as generalized flow networks subject to capacity constraints, which provide a convenient representation of power, bandwidth, and resource limitations (Lattanzi et al., 2007). The maximum sustainable workload (MSW) for a WSN is the so called *maxflow* of the corresponding flow network. Four main results have been recently achieved under this framework (Bogliolo et al., 2010). First, given an EH-WSN and the environmental conditions in which it operates, the theoretical value of the maximum energetically sustainable workload (MESW) can be exactly determined. Second, MESW can be used as a design metric to optimize the deployment of EH-WSNs. Third, the energy efficiency of existing routing algorithms can be evaluated by comparing the actual workload they can sustain with the theoretical value of MESW for the same network. Fourth, self-adapting maxflow (SAMF) routing algorithms have been developed which are able to route the MESW while adapting to time-varying environmental conditions.

This chapter introduces the research field of *design for unlimited lifetime* of EH-WSNs, which aims at exploiting environmental power to maximize the workload of the network under steady-state sustainability constraints. The power harvested at each node is regarded as a time-varying constraint of an optimization problem which is defined and addressed within the theoretical framework of generalized flow networks. The solution of the constrained optimization problem provides the best strategy for managing the network in order to obtain maximum outputs without running out of energy.

The following subsection provides a brief overview of previous work on routing algorithms for autonomous WSNs. Section 2 presents the network model used for analyzing workload sustainability under energy, bandwidth and resource constraints, and introduces the concept of maximum sustainable workload; Section 3 outlines the SAMF routing algorithm, demonstrating its optimality and highlighting its theoretical properties, Section 4 introduces design and simulation tools based on workload sustainability, while Section 5 discusses the practical applicability of SAMF algorithms in light of simulation results obtained by taking into account the effects of non-idealities such as finite propagation time, radio broadcasting, radio channel contention, and packet loss.

1.1 Previous work

The wide range of possible applications and operating environments of *wireless sensor networks* (WSNs) makes scalability and adaptation essential design goals (Dressler, 2008) which have to be achieved while meeting tight constraints usually imposed to sensor nodes in terms of size, cost, and lifetime (Yick et al., 2008). Since the main task of any WSN is to gather data from the environment, the routing algorithm applied to the network is one of the most critical design choices, which has a sizeable impact on power consumption, performance, dependability, scalability, and adaptation. The operation of any WSN usually follows a 2-phase paradigm. In the first phase, called *dissemination*, control information is diffused in order to dynamically change the sampling task (which can be specified in terms of sampling area, target nodes, sampling rate, sensed quantities, ...); in the second phase, called *collection*, sampled data are transmitted from the involved sensor nodes to one or more collection points, called *sinks* (Levis et al., 2008). Routing algorithms have e deep impact on both dissemination and collection phases.

Energy efficiency is a primary concern in the design of routing algorithms for WSNs (Mhatre & Rosenberg, 2005; Shafiullah et al., 2008; Yarvis & Zorzi, 2008). If the routing algorithm requires too many control packets, chooses sub-optimal routes, or requires too many computation at the nodes, it might end up reducing the lifetime of the network because of the limited energy budget of battery-operated sensor nodes. The routing algorithms which have been proposed to maximize network lifetime are documented in many comprehensive surveys (Chen & Yang, 2007; Li et al., 2011; Yick et al., 2008).

Taking a different perspective, lifetime issues can be addressed by means of energy harvesting techniques, which enable the design of autonomous sensor nodes taking their power supply from renewable environmental sources such as sun, light, and wind (Amirtharajah et al., 2005; Nallusamy & Duraiswamy, 2011; Sudevalayam & Kulkarni, 2010). Environmentally-powered systems, however, give rise to additional design challenges due to supply power uncertainty and variability.

While there are a number of routing protocols designed for battery-operated WSNs, only a small number of routing protocols have been published which explicitly account for energy harvesting. *Geographic routing* algorithms (Eu et al., 2010; Zeng et al., 2006) take into account distance information, link qualities, and environmental power at each node in order to select the best candidate region to relay a data packet. Both algorithms, however, strongly depend on the position awareness of sensor nodes, which is difficult to achieve in many WSNs. The environmental power available at each node is used as a weight in the *energy-opportunistic weighted minimum energy* (E-WME) algorithm to determine the weighted minimum path to the sink (Lin et al., 2007).

Moving beyond the opportunity of exploiting environmental power to recharge energy buffers and enhance lifetime, energy harvesting prompt for a paradigm shift from *energy-constrained lifetime maximization* to *power-constrained workload optimization*. In fact, as long as the average workload at each node can be sustained by the average power it takes from the environment, the node can keep working for an unlimited amount of time. In this case rechargeable batteries are still used as energy buffers to compensate for environmental power variations, but their capacity does not affect any longer the lifetime of the network.

It has been shown that autonomous wireless sensor networks can be modeled as flow networks (Bogliolo et al., 2006), and that the *maximum energetically sustainable workload* (MESW) can be determined by solving an instance of *maxflow* (Ford & Fulkerson, 1962). The solution of maxflow induces a MESW-optimal *randomized minimum path recovery time* (R-MPRT) routing algorithm that can be actually implemented to maximally exploit the available power (Lattanzi et al., 2007). Different versions of the R-MPRT algorithm have been proposed to improve performance and reduce packet loss in real-world scenarios, taking into account MAC protocol overhead and lossy wireless channels (Hasenfratz et al., 2010). Environmental changes, however, impose to periodically recompute the global optimum and to update the routing tables of R-MPRT algorithms. A distributed version of maxflow has been proposed that exploits the computational power of WSNs (Kulkarni et al., 2011) to grant to the network the capability of recomputing its own routing tables for adapting to environmental changes (Klopfenstein et al., 2007). Adaptation, however, is a complex task which might conflict with the normal operations of the WSN, thus imposing to trade off adaptation frequency for availability. In general, the adaptation and scalability needs which are typical of WSNs prompt for the application of some sort of self-organization mechanisms (Dressler, 2008; Eu et al., 2010; Mottola & Picco, 2011). In particular, a self-adapting maxflow routing strategy for EH-WSNs has been recently proposed (Bogliolo et al., 2010) which is able to route the maximum sustainable workload under time-varying power, bandwidth, and resource constraints.

2. Network model and workload sustainability

Any WSN can be modeled as a directed graph with vertices associated with network nodes and edges associated with direct links among them: vertices v_i and v_j are connected by an edge $e_{i,j}$ if and only if there is a wireless connection from node i to node j. This chapter focuses on EH-WSNs and retains the symbols introduced by Bogliolo *et al.* (Bogliolo et al., 2010). Each node (say, v) is annotated with two variables: $P(v)$, which represents the environmental power available at that node, and $CPU(v)$, which represents its computational power expressed as the number of packets that can be processed in a time unit. Similarly, each edge (say, e), is annotated with variable $C(e)$, which represents the capacity (or bandwidth) of the link, and variable $E(v, e)$, which represents the energy required at node v to process (receive or generate) a data packet and to forward it through its outgoing edge e.

The maximum number of packets that can be steadily sent across e in a time unit (denoted by $cap(e)$) is limited by its bandwidth ($C(e)$), by the processing speed of the source node ($CPU(v)$), and by the ratio between the environmental power available at the node and the energy needed to process and send a packet across e ($P(v)/E(v, e)$). In fact, the ratio between the energy needed to process a packet and the power harvested from the environment represents the time required to recharge the energy buffer in order to be ready to process a new packet. The inverse ratio is an upper bound for the sustainable packet rate. In symbols:

$$F(e) \leq cap(e) = \min\{C(e), CPU(v), \frac{P(v)}{E(v, e)}\} \tag{1}$$

where $F(e)$ is the packet flow over edge e. Since $cap(e)$ is an upper bound for $F(e)$, it can be treated as a link capacity that summarizes all the constraints applied to the edge, suggesting that the overall sensor network can be modeled as a *flow network* (Ford & Fulkerson, 1962).

Each node, however, usually has multiple outgoing edges that share the same power and computational budget, so that capacity constraints cannot be independently associated with the edges without taking into account the additional constraints imposed to their source nodes, represented by the following equations:

$$\sum_{e\_exiting\_from\_v} F(e) \leq CPU(v) \tag{2}$$

$$\sum_{e\_exiting\_from\_v} F(e)E(v, e) \leq P(v) \tag{3}$$

If the transmission power is not dynamically adapted to the actual length of the wireless link (Wang & Sodini, 2006), the energy spent at node v to process a packet can be regarded as a property of the node (denoted by $E(v)$) independent of the outgoing edge of choice. In this case, which is typical of most real-world WSNs, the constraints imposed by Equations 2 and 3 can be suitably expressed as capacity constraints (denoted by $cap(v)$) applied to the packet flow across node v (denoted by $F(v)$). In symbols:

$$F(v) = \sum_{e\_exiting\_from\_v} F(e) \tag{4}$$

$$F(v) \leq cap(v) = \min\{CPU(v), \frac{P(v)}{E(v)}\} \tag{5}$$

Node-constrained flow networks can be easily transformed into equivalent edge-constrained flow networks by splitting each original constrained node (v) into an *input sub-node* (destination of all incoming edges) and an *output sub-node* (source of all outgoing edges) connected by an internal (virtual) edge with capacity $cap(v)$ (Ford & Fulkerson, 1962). All other edges, representing the actual links among the nodes, maintain their original capacities according to Equation 1. The result is an edge-constrained flow network which retains all the constraints imposed to the sensor network, allowing us to handle EH-WSNs within the framework of flow networks.

When node constraints cannot be expressed as cumulative flow limitations independent of the incoming or outgoing edges, however, the network cannot be transformed into an equivalent edge-constrained flow network. Any directed graph with arbitrary flow limitations possibly imposed at both edges and nodes, will be hereafter called *generalized flow network*.

For the sake of explanation we consider sensor networks made of 4 types of nodes: *sensors*, which are equipped with transducers that make them able to sense the environment and to generate data packets to be sent to a collection point, *sinks*, which generate control packets and collect data packets, *routers*, which relay packets according to a given routing algorithm, and *sensor-routers*, which exhibit the behavior of both sensors and routers. Without loss of generality, in the following we consider a sensor network with only one sink. Generalization to multi-sink networks can be simply obtained by adding a dummy sink connected at no cost with all the actual sinks (modeled as routers).

Figure 1 shows a hierarchical sensor network (Iwanicki & van Steen, 2009) of 64 sensors (thin circles), 16 routers (thick circles), and 1 sink (square) which will be used throughout the rest of this chapter to illustrate the routing strategy and to test its performance. Sensors and routers are uniformly distributed over a square 10x10 region, with the sink in the middle. The communication range of each node is equal to the minimum diagonal distance between the routers (edges are not represented for the sake of simplicity). Shading is used to highlight the sensors that need to be sampled according to a given monitoring task. The case of Figure 1 refers to a monitoring task involving only the 4 sensor nodes in the upper-left corner of the coverage area.

Definition 1. Given a WSN, the environmental conditions (expressed by the distribution of environmental power available at each node), and a monitoring task, the *maximum sustainable workload* (MSW) for the network is the maxflow from the sampled sensors to the sink in the corresponding flow network.

Since maxflow is defined from a single source to a single destination (Ford & Fulkerson, 1962), if there are multiple sensors that generate packets simultaneously, a dummy source node with cost-less links to the actual sources needs to be added to the model. The maxflow from the dummy source to the sink represents the global MSW.

The MSW can be determined in polynomial time by solving an instance of the *maximum-flow problem* within the theoretical framework of flow networks (Ford & Fulkerson, 1962).

If the transmission power is tuned to the length (and quality) of the links, the energy per packet depends on the outgoing edge, so that Equations 2 and 3 cannot reduce to Equation 5, node constraints cannot be transformed into equivalent edge constraints, and classical maxflow algorithms cannot be applied. Nevertheless, the network is still a generalized flow network, the maxflow of which represents the MSW of the corresponding WSN.

The theory presented in this chapter is not aimed at determining the MSW of a WSN. Rather, it is aimed at designing a routing algorithm able to route any sustainable workload, including the theoretical maximum. Hence, we are interested in the value of MSW at the only purpose of testing the routing algorithm under worst case operating conditions.

If the monitoring task consists of sampling a given subset of the sensor nodes, the MSW is directly related to the *maximum sustainable sampling rate* (MSSR) at which all target nodes can be simultaneously sampled by the sink without violating power, bandwidth, and resource constraints (Lattanzi et al., 2007).

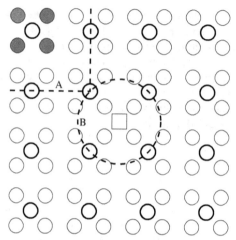

Fig. 1. Hierarchical network used as a case study.

3. Self-adapting maxflow routing algorithm

Capacity constraints, path capacities, and bandwidth requirements can be expressed in terms of packets per time unit. Since dynamic routing strategies can take different decisions for routing each packet, the routing algorithm can be developed by looking at packets rather than at overall flows. The capacity constraints imposed at a given node (edge) at the beginning of a time unit represent the actual number of packets that can be processed by that node (routed across that edge) in the time unit. Whenever the node (edge) is traversed by a packet, its *residual capacity* (which represents its capability of handling other packets in the same time unit) is decreased because of the energy, CPU, and bandwidth spent to process that packet.

Given a generalized flow network with vertices V and edges E, a path of length n from a source node s to a destination node d is a sequence of nodes $\mathcal{P} = (v_0, v_1, ..., v_{n-1})$ such that $v_0 = s$, $v_{n-1} = d$, and $e_{v_{i-1},v_i} \in E$ for each $i \in [1, n-1]$. We call *path capacity* of $\mathcal{P}$, denoted by $cap(\mathcal{P})$, the maximum number of packets per time unit that can be routed across the path without violating any node or edge constraint. We call *point-to-point flow* the flow of packets routed from one single source to one single destination, regardless of the path they follow.

Referring to a path $\mathcal{P}$ and to a time unit t, the *nominal path capacity* of $\mathcal{P}$ at time t is the capacity of the path computed at the beginning of the time unit by assuming that all the resources along the path are entirely assigned to $\mathcal{P}$ for the whole time unit. In practice, it corresponds to the minimum of the capacities of the nodes and edges belonging to the path, as computed at the beginning of the time unit. The *residual path capacity* of $\mathcal{P}$ at time $t + \tau$, on the contrary, is the path capacity re-computed at time $t + \tau$ by taking into account the resources consumed up to that time by the packets processed since the beginning of the time unit.

The *self-adapting maxflow* (SAMF) routing algorithm proposed by Bogliolo *et al.* (Bogliolo et al., 2010) implements a simple *greedy* strategy that can be described as follows: *always route packets across the path with maximum residual capacity to the sink.*

According to this strategy, the residual path capacity is used as a routing metric. More precisely, the metric used at node v to evaluate its outgoing edge e is the maximum of the residual capacities of all the paths leading from v to the sink through edge e. The minimum number of hops can be used as a second criterion to choose among edges with the same residual path capacity.

The complexity of the routing algorithm is hidden behind the real-time computation of residual path capacities, which are possibly affected by any routed packet and by any change in the constraints imposed to the nodes and to the edges encountered along the path. In principle, in fact, routing metrics should be recomputed at each node (and possibly diffused) whenever a data packet is processed or an environmental change is detected.

In order to reduce the control overhead of real-time computation of residual path capacities, routing metrics can be kept unchanged for a given time period (called *epoch*) regardless of traffic conditions and environmental changes, and recomputed only at the beginning of a new epoch. In this way, all the packets processed by a node (say v) in a given epoch are routed along the same path, which is the one with the highest nominal capacity as computed at the beginning of that epoch. Residual capacities are computed at the end of the epoch by subtracting from nominal capacities the cost of all the packets routed in that epoch (in terms of energy, CPU, and bandwidth).

The lack of feedback on the effects of the routing decisions taken within the same epoch may cause the nodes to keep routing packets along saturated paths, leading to negative residual capacities at the end of the epoch. Negative residual capacities (hereafter called *capacity debts*) represent temporary violations of some of the constraints. Depending on the nature of the constraints (power, CPU, bandwidth) the excess flow that causes a capacity debt can be interpreted either as the amount of packets enqueued at some node waiting for the physical resources (bandwidth or CPU) needed to process them, or as the extra energy taken at some node from an auxiliary battery that needs to be recharged in the next epoch. In any case, capacity debts need to be compensated in subsequent epochs. This is done by subtracting the debts from the corresponding nominal capacities before computing nominal path capacities at the beginning of next epoch. **Example 1.** Consider the network of Figure 1 with the same constraints (namely, $cap(v) = 200$) imposed to all sensors and routers, and with no edge constraints. The effect of a SAMF routing strategy are shown in Figure 2, where sensor nodes and edges not involved in the monitoring task are not represented for the sake of simplicity. Intuitively, the maxflow is 600, corresponding to a MSSR of 150 packets per sensor per unit. In fact, all data packets need to be routed across cut A (shown in Figure 1), which contains only 3 routers with an overall capacity of 200x3=600 packets per time unit. An optimal flow distribution is shown in Figure 2.d, where the thickness of each edge represents the flow it sustains: 50 packets per time unit for the thin lines, 200 packets per time unit for the thick

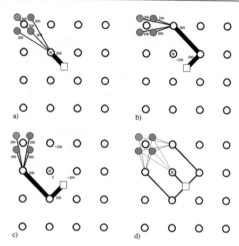

Fig. 2. d) Maxflow of the example network, obtained by averaging the flows over three epochs: a), b), and c).

ones. Figures 2.a, 2.b, and 2.c show the flows allocated by the SAMF routing strategy in three subsequent epochs (of one time unit each) when the 4 sensors of interest are sampled at the MSSR (namely, 150 packets per time unit). Nominal node capacities at the beginning of each epoch are annotated in the graphs in order to point out the effects of over-allocation. In the first epoch all the paths to the sink exhibit the same capacity, so that the path length is used to choose the best path. Since the routing metric is not updated during the first epoch, all data packets (namely, 600) are pushed along the shortest path from the upper corner to the sink, which could sustain only 200. The residual capacity of node n at the end of the first epoch is -400. The capacity debt of 400 packets is then subtracted from the nominal capacity of node n (200) at the beginning of the second epoch, that becomes -200. This negative value imposes to the algorithm the choice of a different path. Since the capacity debt of the shortest path is completely compensated at the end of the third epoch, the entire routing strategy can be periodically applied every three epochs. The optimal maxflow solution of Figure 2.d can be obtained by averaging the flows allocated by the self-adapting algorithm in the three epochs shown in the figure.

We say that a routing strategy *converges* if it can run forever causing only finite capacity debts. The convergence property of the SAMF routing strategy is stated by the following theorem.

Theorem 1. Given an autonomous WSN with power, bandwidth, and resource constraints expressed by Equations 1 and 5, the SAMF routing strategy converges for any sustainable workload.

Proof. Assume, by contradiction, that a sustainable workload is applied to the network but the strategy does not converge, so that there is at least one edge (or one node) with a capacity debt which keeps increasing and a residual capacity which decreases accordingly. Without loss of generality, assume that edge e, from node i to node j, has the lowest residual capacity at the end of time epoch h, denoted by $cap(e)^{(h)}$. If the routing strategy does not converge, for each epoch h there is an epoch $k > h$ such that the residual capacity of e at the end of k is lower than that at the end of h. In symbols:

$$cap(e)^{(k)} < cap(e)^{(h)}$$

The decrease of the residual capacity in e from epoch h to epoch k means that the average flow routed across e in that interval of time exceeds the nominal capacity of e. We distinguish two different cases.

In the first case all the point-to-point flows routed across e have no alternate paths to the sink. Hence, if the sum of the flows exceeds the capacity of e, then the workload is not sustainable. A contradiction.

In the second case the edge is also used to route at least one multi-path point-to-point flow. Since the routing metric is based on residual path capacities, a path including edge e can be taken iff all alternate paths have lower residual capacities. Since e was the edge with lowest residual capacity at epoch h, then it can be taken at some epoch l from h to k iff the capacities of all alternate paths have become lower in the mean time. But this may happen iff the average flows routed across all alternate paths have exceeded their nominal capacities, meaning that the workload is not sustainable for the network. A contradiction.

Theorem 1 demonstrates that SAMF routing is able to route any sustainable workload under power, bandwidth, and resource constraints. Moreover, it has the inherent capability of adapting to environmental changes expressed as time-varying constraints.

It is worth noticing, however, that the theory exposed so far doesn't take into account the control traffic overhead required to recompute routing metrics at the beginning of each time epoch, nor the non-idealities of the wireless channels used for communication. The impact of traffic overhead and non idealities will be extensively discussed in Section 5.

4. Design and simulation tools

The theoretical framework described in Section 2 and the routing algorithm outlined in Section 3 provide the basis for the development of design methodologies for WSNs with unlimited lifetime.

In fact, the algorithmic solutions to the maximum-flow problem (Ford & Fulkerson, 1962) enable the evaluation of the MSW of a given WSN under specific environmental conditions and monitoring tasks. Techniques for the computation of the *maximum energetically sustainable workload* (MESW) were proposed by Bogliolo *et al.* (Bogliolo et al., 2006) for different monitoring tasks, including *selective monitoring* (i.e., sampling of a single node at the maximum sustainable rate), *non-uniform monitoring* (i.e., sampling a cluster of sensor nodes to generate the maximum overall traffic), and *uniform monitoring* (i.e., sampling a cluster of nodes at the maximum sustainable common rate). While the first two tasks can be directly solved as instances of maxflow, uniform monitoring requires an iterative approach which makes use of maxflow in the inner loop (Bogliolo et al., 2006). The same algorithms originally developed to determine MESW, can be applied to determine the more general MSW, which also takes into account CPU and capacity constraints.

The capability of evaluating the MSW can be used, in its turn, to drive the design of sustainable routing algorithms and the deployment of energy-aware WSNs. The concept of *MESW optimality* was introduced to this purpose (Lattanzi et al., 2007). A MESW-optimal non-deterministic routing algorithm can be directly derived from the solution of maxflow: each edge can be chosen by the algorithm with a probability proportional to the amount of flow across that edge in the solution of maxflow. Edge probabilities can be stored as routing tables at each node and used at run time to take pseudo-random decisions. A static version of this algorithm was originally implemented on *Tmote Sky* nodes (Klopfenstein et al., 2007). The SAMF routing algorithm outlined in Section 3 achieves the same MESW optimality while dynamically adapting to time-varying conditions.

The self-adapting capabilities of SAMF routing, together with the proof of optimality given by Theorem 1, provide a practical mean for overcoming the limitation of maxflow algorithms which cannot be applied to generalized flow networks subject to node constraints that cannot be transformed into edge constraints. First of all, thanks to the proved optimality, SAMF algorithm can be directly applied to any WSN without further optimization steps. Second, a WSN running the SAMF algorithm can be used to check the sustainability of a given workload. Iterative approaches have been developed on top of a simulation model of SAMF algorithm to determine the MSW of generalized flow networks (Seraghiti et al., 2008), as detailed in Subsection 4.2.

Finally, accurate network simulation models are required to investigate the practical applicability of MSW-optimal algorithms by evaluating the additional features of practical interest (such as control traffic overhead, number of hops, convergence speed, maximum buffer size, and scalability) and the effects of real-world non-idealities (such as communication time, radio broadcasting, radio-channel contention, and packet loss). The inherent features of the SAMF algorithm were evaluated by running extensive experiments with the simulation model implemented on top of OMNeT++, a discrete-event, open-source, modular network simulator (Bogliolo et al., 2010). Network components were written in C++ and composed using a high-level network description language called *NED*. The evaluation of the impact of non-idealities has prompted for the development of a more realistic simulation model which is presented in the following subsection.

4.1 SAMF simulator

The simulator presented in this subsection has been conceived to allow the designer to directly execute the Java bytecode written for the target sensor nodes (namely, `Sentilla JCreate` or other sensor nodes running an embedded JVM) while simulating their power consumption and the effects of realistic channel models. Since non-idealities can be selectively enabled or turned off, the simulator bridges the gap between theory and practice in that it can be used both to reproduce the theoretical results (when launched with ideal channel models) and to simulate real-world conditions (when launched with channel and energy models characterized on the field).

The simulator has been developed on top of the `SimJava` framework (Kreutzer et al., 1997), an event-driven multi-threaded simulator which handles concurrent entities (`Sim_entity` objects) running on separate threads and communicating through uni-directional channels established between their ports. The multi-threaded nature of SimJava has been deeply exploited to simulate the concurrency among the nodes and among the tasks executed at each node. Each sensor node has been implemented by means of two separate instances of Sim_entity: a SensorSw executing the bytecode to be loaded on the target sensor node and a SensorHw catching low level calls and emulating the behavior of the hardware, including the routing protocol. In case of multiple applications running on the same node, each of them is assigned an independent instance of SensorSw.

In order to make it possible for the simulator to run the same bytecode compiled for the target Sentilla nodes, the libraries included in the Sentilla framework have been re-implemented by exposing the same API while allowing the SensorHw to: catch the appropriate calls, exhibit the required behavior, emulate hardware devices (including LED, radio devices, and sensors), and enable instrumentation.

Modeling wireless (i.e., broadcast) communication channels on top of SimJava has required the development of new class (i.e., Network) which extends Sim_Entity and works as a network dispatcher. In practice, it is connected to all sensor nodes through bi-directional virtual channels (with no correspondences with real-world channels) which allow the network to take full control of the actual topology of the WSN, to catch all communication events, to implement channel models, to inject non-idealities, and to deliver packets.

The Network class does not contain the channel model. Rather, the model is specified in a separate object, called ChannelBehavior, which is loaded by the constructor. The level of realism of the simulation can be tuned by changing the channel behavior.

4.2 Test of sustainability

According to Theorem 1, the SAMF routing strategy implemented by the simulation model outlined so far is guaranteed to converge for any sustainable workload. Hence, simulation stability can be used as a proof of sustainability for the workload applied to the network. The instability of the simulation can be detected by checking for capacity debts that keep increasing over time. The simulation is considered to be convergent if it lasts for a "long-enough" number of epochs without causing any instability.

Given the simulation model of a WSN, the simulator can be viewed as a function which takes in input two parameters: the sampling rate (SR) to be applied to the target sensors and the simulation length (Nepochs). When the function is invoked the simulation is launched and the function returns 1 in case of instability or 0 in case of normal termination. The function can then be used within the inner loop of a bisection-search algorithm in order to estimate the *maximum sustainable sampling rate* of a uniform monitoring task. We denote by $M\tilde{S}SR$ the estimated value of MSSR, which suffers from two sources of approximation: the limited number of iterations in the bisection-search algorithm (each iteration adds a binary digit to the precision) and the limited length of simulation runs (which avoids the detection of instabilities that would show up after the end of simulation). As a result, $M\tilde{S}SR$ overestimates the actual value of MSSR with a precision given by $n \log_{10} 2$, where n is the number of bisection iterations. For instance, if the bisection algorithm iterates $n = 10$ times, then the precision of the estimator is 3 digits, corresponding to a maximum overestimation of 1 per mil.

5. Simulation results

This section makes use of the simulator described in Section 4.1 in order to test and discuss the performance of the SAMF algorithm and its sensitivity to design parameters and real-world operating conditions. The parametric nature of the simulator and its capability of using different channel models are used hereafter to validate simulation results against their theoretical counterparts (Subsection 5.1), to incrementally add non-idealities to make simulation more realistic (Subsection 5.2), and to conduct a sensitivity analysis on a large set of Monte Carlo experiments (Subsection 5.3).

5.1 Validation

Validation was performed by running the simulation-based iterative procedure outlined in Section 4.2 with ideal channel models in order to determine the value of $M\tilde{S}SR$ (i.e., the estimated value of the maximum sustainable sampling rate). WSNs without edge-dependent node constraints were used to this purpose in order to apply classical maxflow algorithms (Ford & Fulkerson, 1962) to compute the theoretical value of $MSSR$ on the equivalent flow network. The ratio between $M\tilde{S}SR$ and $MSSR$ represents the so-called *optimality ratio*, which was originally introduced to express the optimality of routing algorithms (Lattanzi et al., 2007). When the optimality of the algorithm under study is known a priori, as in case of the SAMF algorithm applied in ideal conditions, the ratio provides a measure of the accuracy of the simulator.

The validation procedure was applied to the example of Figure 1 by running the bisection search algorithm for 20 iterations with simulations lasting for 1,000 epochs of 50 time units each. The optimality ratio obtained whitouth taking into account the control traffic overhead (in order to make simulation results directly comparable with the theoretical optimum) was 0.990, while the value achieved with control traffic overhead was 0.973, demonstrating both the accuracy of the simulator and the small overhead of control packets in the experimental settings adopted.

5.2 Effects of non-idealities

Specific models were implemented on top of the simulator described in Section 4.1 to investigate the effects of the following non-idealities:

- *communication delay*, which represents the propagation time across the channel;
- *de-synchronization*, which is modeled as a boot time randomly generated for each node;
- *transmission time*, which represents the time required by the transmitter node to send the whole packet across the channel;
- *channel collision*, which avoids a destination node to properly receive two packets with overlapping transmission times;
- *packet loss*, which represents the probability of discarding a packet because of the bit error rate of the channel;
- *reception energy*, which represents the energy spent to receive a packet at each node which is in the range of the transmitter, independently of the destination address.

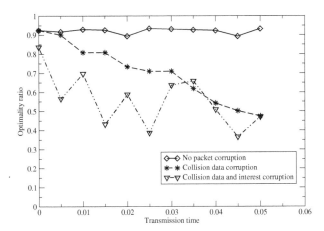

Fig. 3. Optimality ratio as a function of transmission time.

The results presented hereafter refer to the simple WSN of Figure 1 simulated with the same parameters used for validation. It is worth noting that the MSW is measured as the overall number of packets received at the sink in a time unit when sampling at the same rate the 4 sensor nodes placed in the upper-left corner of the coverage area.

5.2.1 Timings

The introduction of communication delay and de-synchronization produced negligible effects on the optimality ratio, demonstrating the robustness of the SAMF algorithm with respect to timing uncertainties and misalignments. Simulating a non-null transmission time has the only effect of keeping the channel busy during transmission, imposing an upper bound to the packet rate. If the simulated production rate does not exceed the physical upper bound of the channel, transmission time does not impact simulation results. This is shown by the solid curve of Figure 3, which plots the optimality ratio as a function of transmission time.

5.2.2 Channel contention

The time spent to send a packet has a sizeable impact on performance when channel collision is simulated. In this case, in fact, the channel cannot be simultaneously used by neighboring nodes, or otherwise collisions would cause the corruption of the packets. Channel sensing mechanisms with pseudorandom retry time were simulated in order to manage channel contention. Although this simple mechanism does not avoid collisions occurring at a destination node receiving simultaneous packets from two or more non adjacent nodes, it affects the optimality ratio for three main reasons: first, because of the loss of collided packets which do not reach the sink, second, because of the induced correlation between the activity of adjacent nodes, third, because of the reduced path diversity of the SAMF algorithm. In fact, since collided packets are discarded at the point of collision, they do not consume any energy along the rest of the path to the sink. Hence, the routing metrics based on residual energy induce the routing algorithm to keep sending packets across high-collision paths. The combined effect of these three phenomena is shown by the decreasing trend of the dashed

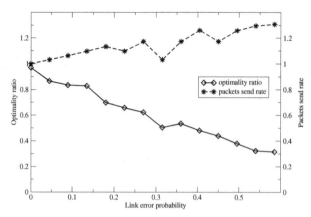

Fig. 4. Optimality ratio as a function of link error probability.

curve of Figure 3. The chaotic behavior of the dash-dotted curve is obtained by simulating also collisions occurring in the diffusion phase, thus causing a loss of interest packets which ultimately impacts the correctness of the SAMF algorithm.

5.2.3 Link quality

Figure 4 plots the effects of packet loss due to link error probability independent of packet collisions. Although the MSW decreases for increasing values of the link error probability, the effect is much easier to explain than that caused by packet collisions. In this case, in fact, packet loss is independent of traffic congestion and it does not induce any correlation between adjacent nodes. Hence, the loss of optimality is only caused by the reduced percentage of packets which reach the sink. The dashed curve in Figure 4 shows the increased sampling rate imposed to the sensor nodes in order to compensate for the loss of packets. A deeper analysis of simulation results highlights that the higher the link error rate the shorter the paths used on average to route the packets. In fact, since the errors are independently injected at each link, the probability of losing a packet along a path increases with the path length.

5.2.4 Reception energy

Wireless transmission is based on radio broadcasting. This means that each node receives all the packets transmitted by its adjacent nodes, even if it is not along the path selected by the routing algorithm. In case of point-to-point transmission across a wireless link, the packet is discarded by all the receiving nodes but the destination one. However, some energy is spent at each node to receive the packet and check its destination address before taking the decision to discard it. The energy wasted to listen to a broadcast channel has a deep impact on the energy efficiency of a WSN. This phenomenon is often neglected by energy-aware routing algorithms, which are mainly focused on transmission/processing energy spent by nodes which lay along the routing path. Figure 5 plots the optimality ratio as a function of the ratio between the reception (RX) and transmission (TX) energy of the sensor nodes. It is worth noticing that when RX energy is about one tenth of TX energy, the MSW reduces to 50% of the theoretical optimum. When RX energy equals TX energy (which is a typical situation) the optimality

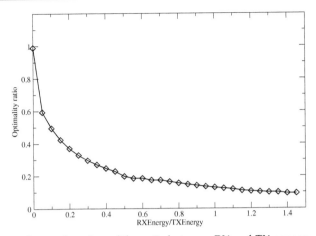

Fig. 5. Optimality ratio as a function of the ratio between RX and TX energy.

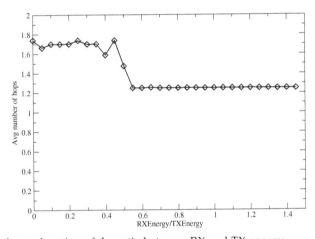

Fig. 6. Path length as a function of the ratio between RX and TX energy.

ratio reduces below 20%. Figure 6 reports the average number of hops from the source sensor nodes to the sink, as a function of RX energy. Since the packets routed along the best path also cause a sizeable waste of energy in the neighboring paths, RX energy significantly reduces the degrees of freedom available to the SAMF algorithm. For the case study of Figure 1, when RX energy accounts for more than 60% of TX energy, the crosstalk effect avoids the SAMF algorithm to take advantage of path diversity and the routing strategy resorts to minimum path.

5.3 Monte Carlo experiments

Monte Carlo simulations were conducted to perform a sensitivity analysis by means of pseudo-random sampling in a neighborhood of a given point in the design space. To this purpose, we used parametrized randomly generated WSNs composed of nodes scattered in a square region with a sink in the middle. All the nodes (but the sink) have sensing and routing capabilities and are involved on a uniform-sampling task targeting the whole area.

The following simulation parameters were used as independent variables: the number of sensor nodes ($NSensors$), the transmission range of each node ($TxRange$), the energy spent at each node to transmit and process a packet ($TxEnergy$), the energy spent at each node to receive a packet ($RxEnergy$), the environmental power available at each node ($Power$), the length of the time epochs adopted for recomputing routing metrics ($Epoch\ length$), the capacity of the energy buffer installed at each node ($Energy\ buffer$), the transmission time ($TxTime$), the propagation delay ($Link\ delay$), and the link error probability ($Link\ err.\ prob.$).

The sensitivity analysis was conducted for the following (dependent) parameters of interest: the estimated values of MSW ($M\tilde{S}W$), the maximum capacity debt observed during the whole simulation ($MDebt$), the average path length ($Path\ l.$), the control traffic overhead ($Overhead$), the total amount of data packets routed ($Routed\ data$), and the total amount of collisions occurred during simulation ($Collision$).

The analysis was conducted on a sample of 1,000 points in parameter space. Each sampling point corresponds to a configuration of the independent variables uniformly taken from the ranges reported in Table 1. For each configuration, 3 random *trials* were performed using different seeds, resulting in 3,000 runs of the simulation-based bisection search of $M\tilde{S}W$.

Table 1 summarizes the results of the sensitivity analysis. Rows and columns are associated with independent variables and dependent parameters, respectively. The second column reports the sampled value range of each independent variable, while the second and third rows report the sample average and standard deviation of the dependent parameters. All other entries of Table 1 report the correlation coefficients between independent and dependent variables computed on the results of the 3,000 Monte Carlo experiments. The most significant correlations (with absolute value greater or equal than 0.2) are highlighted in boldface and discussed in the following, column by column.

- $M\tilde{S}W$ is negatively affected by the number of sensor nodes ($NSensors$), which reduces the sustainable sampling rate of each sensor because of the limited routing capabilities of the network, and positively affected by the transmission range ($TxRange$), which increases the number of paths available to route data packets. Interestingly enough, the negative impact of $RxEnergy$ on $M\tilde{S}W$ is much higher than that of $TxEnergy$, because of the energy waste induced in the neighborhood of the routing path discussed in Section 5.2. As expected, a high correlation coefficient is observed between $M\tilde{S}W$ and environmental power ($Power$), while the link error probability negatively affects the maximum sustainable workload.

- $Mdebt$ is mainly affected by the environmental power because of its high correlation with $M\tilde{S}W$. In fact, the debt is caused by the excess of packets routed across a saturated path in a time epoch because of the lack of feedback on the residual path capacity. The higher the sampling rate, the higher the debt that can be reached during simulation.

- $Path\ length$ increases with the overall flow, which depends, in its turn, from the environmental power ($Power$). In fact, the larger the flow the larger the number of paths (possibly longer than the minimum one) that need to be used by the routing

		$M\tilde{S}W$	$MDebt$	Path l.	Overhead	Routed data	Collisions
	Average	0.18	25,095	2.688	20,427	24,934	19,729
	STD	0.09	11,753	1.302	11,003	10,946	28,609
Edge	[100, 200]	0.181	-0.027	0.117	-0.028	0.052	-0.096
NSensors	[15, 25]	**-0.218**	0.040	-0.084	**0.541**	-0.069	0.187
TxRange	[50, 100]	**0.399**	0.167	**-0.779**	0.243	**-0.725**	-0.062
TxEnergy	[100, 200]	-0.140	-0.048	-0.005	0.036	0.011	-0.035
RxEnergy	[100, 200]	**-0.405**	0.011	0.026	0.031	0.050	**-0.308**
Power	[500, 5000]	**0.433**	**0.492**	**0.239**	-0.022	-0.039	**0.265**
Epoch length	[50, 100]	-0.030	-0.004	-0.012	**-0.688**	-0.011	-0.049
Energy buffer	[50k, 500k]	0.024	0.004	-0.031	0.002	0.046	0.011
TxTime	[0.0, 0.05]	-0.012	-0.067	0.065	-0.097	0.047	**0.672**
Link delay	[0.0, 0.1]	0.004	0.102	-0.043	-0.082	-0.073	0.133
Link err. prob.	[0.0, 0.5]	**-0.374**	0.056	**-0.271**	0.147	**-0.211**	0.087

Table 1. Results of the sensitivity analysis, expressed by the correlation coefficients between independent variables (rows) and dependent parameters (columns). Significant correlations are highlighted in bold.

strategy. As expected, the average path reduces for longer transmission ranges (TxRange). Interestingly, path length decreases for higher values of link error probability (Link err. prob.). This is due to the fact that statistics are computed on packets received by the sink, and packets routed across longer paths have a lower probability of reaching the sink.

- Overhead is positively affected by the number of nodes (NSensors) and by the transmission range (TxRange). In fact, the number of Interest messages received and sent by each sensor depends on the number of incoming and outgoing edges, respectively. Both the number of nodes and the transmission range positively affect the degree of connectivity, causing a larger overhead. As already discussed, the overhead reduces when the epoch length increases.

- Routed data is negatively affected by the transmission range (TxRange), which decreases the number of hops needed to reach the sink, and by link error probability (Link err. prob.), which favors shorter paths.

- Collisions is positively affected by Power and negatively affected by RxEnergy. Both correlations can be explained by looking at the maximum sustainable workload. In fact, the collision probability increases with traffic. Finally, collisions are strongly affected by the transmission time, which increases the risk of overlapping of the transmission intervals of two or more independent packets sent to the same node.

6. Conclusions

Energy harvesting (EH) techniques enable the development of wireless sensor networks (WSNs) with unlimited lifetime. This attractive perspective prompts for a paradigm shift in the design and management of EH-WSNs.

This chapter has provided a thorough overview of the results recently achieved in the design of EH-WSN within the framework of generalized flow networks, including the network model, the concept of maximum sustainable workload (MSW), and the self-adapting maxflow routing algorithm (SAMF). In particular, the SAMF algorithm is able to route any theoretically sustainable workload while autonomously adapting to time varying environmental conditions. It has been shown that the MSW is a suitable design metric for EH-WSNs with unlimited lifetime, while the SAMF algorithm can be used within the inner loop of bisection search algorithms to estimate the MSW for generalized flow networks which cannot be handled by traditional maxflow algorithms.

Finally, a new simulator has been developed to evaluate the practical applicability of the theoretical results. The simulator has been validated by reproducing theoretical results under ideal operating conditions, and then used to inject real-world non idealities, including propagation delay, de-synchronization, channel contention, packet loss, and reception energy. The sensitivity analysis conducted on a large set of Monte Carlo simulation experiments allows the designer to figure out the performance of the SAMF algorithm in many different real-world scenarios. In particular, it has been pointed out that the maximum sustainable workload is highly affected by the reception energy which is spent at each node to receive broadcast packets independently of their destination address. The reception energy wasted by nodes which are not along the routing path is usually neglected by energy-aware routing algorithms since it is a side effect which is not captured by routing metrics.

Future directions in the field of WSNs with unlimited lifetime include: modeling reception energy within the framework of generalized flow networks, developing sensor nodes able to reduce the energy wasted in listening for packets addressed to other nodes, developing design tools using MSW as a metric, and implementing SAMF routing strategies in real-world WSNs.

7. References

Amirtharajah, R., Collier, J., Siebert, J., Zhou, B. & Chandrakasan, A. (2005). Dsps for energy harvesting sensors: applications and architectures, *IEEE Pervasive Computing* 4(3): 72–79.

Bogliolo, A., Delpriori, S., Lattanzi, E. & Seraghiti, A. (2010). Self-adapting maximum flow routing for autonomous wireless sensor networks, '*Cluster Computing* 14: 1–14.

Bogliolo, A., Lattanzi, E. & Acquaviva, A. (2006). Energetic sustainability of environmentally powered wireless sensor networks, *Proceedings of the 3rd ACM international workshop on Performance evaluation of wireless ad hoc, sensor and ubiquitous networks*, pp. 149–152.

Chen, H.-H. & Yang, Y. (2007). Guest editorial: Network coverage and routing schemes for wireless sensor networks, *Elsevier Computer Communications* 30(14-15): 2697–2698.

Dressler, F. (2008). A study of self-organization mechanisms in ad hoc and sensor networks, *Elsevier Computer Communications* 31: 3018–3029.

Eu, Z. A., Tan, H.-P. & Seah, W. K. (2010). Opportunistic routing in wireless sensor networks powered by ambient energy harvesting, *Computer Networks* 54(17): 2943 – 2966.

Ford, L. R. & Fulkerson, D. R. (1962). *Flows in Networks*, Princeton University Press.

Hasenfratz, D., Meier, A., Moser, C., Chen, J.-J. & Thiele, L. (2010). Analysis, comparison, and optimization of routing protocols for energy harvesting wireless sensor networks, *Proceedings of Sensor Networks, Ubiquitous, and Trustworthy Computing*, pp. 19–26.

Iwanicki, K. & van Steen, M. (2009). On hierarchical routing in wireless sensor networks, *Proceedings of the 2009 International Conference on Information Processing in Sensor Networks*, pp. 133–144.

Klopfenstein, L. C., Lattanzi, E. & Bogliolo, A. (2007). Implementing energetically sustainable routing algorithms for autonomous wsns, *International Symposium on a World of Wireless, Mobile and Multimedia Networks (WoWMoM 2007)*, pp. 1–6.

Kreutzer, W., Hopkins, J. & van Mierlo, M. (1997). Simjava a framework for modeling queueing networks in java, *Proceedings of the 29th conference on Winter simulation*, pp. 483–488.

Kulkarni, R., Forster, A. & Venayagamoorthy, G. (2011). Computational intelligence in wireless sensor networks: A survey, *Communications Surveys and Tutorials, IEEE* 13(1): 68 – 96.

Lattanzi, E., Regini, E., Acquaviva, A. & Bogliolo, A. (2007). Energetic sustainability of routing algorithms for energy-harvesting wireless sensor networks, *Elsevier Computer Communications* 30(14-15): 2976–2986.

Levis, P., Culler, D., Gay, D., Madden, S., Patel, N., Polastre, J., Shenker, S., Szewczyk, R. & Woo, A. (2008). The emergence of a networking primitive in wireless sensor networks, *Communications of the ACM* 51(7): 99–106.

Li, C., Zhang, H., Hao, B. & Li, J. (2011). A survey on routing protocols for large-scale wireless sensor networks, *Sensors* 11(4): 3498–3526.

Lin, L., Shroff, N. B. & Srikant, R. (2007). Asymptotically optimal energy-aware routing for multihop wireless networks with renewable energy sources, *IEEE/ACM Trans. Netw.* 15: 1021–1034.

Mhatre, V. & Rosenberg, C. (2005). Energy and cost optimizations in wireless sensor networks: A survey, *in* A. Girard, B. Sanso & F. Vazquez-Abad (eds), *Performance Evaluation and Planning Methods for the Next Generation Internet*, Kluwer Academic Publishers.

Mottola, L. & Picco, G. P. (2011). Programming wireless sensor networks: Fundamental concepts and state of the art, *ACM Comput. Surv.* 43: 19:1–19:51.

Nallusamy, R. & Duraiswamy, K. (2011). Solar powered wireless sensor networks for environmental applications with energy efficient routing concepts: A review, *Information Technology Journal* pp. 1–10.

Seraghiti, A., Delpriori, S., Lattanzi, E. & Bogliolo, A. (2008). Self-adapting maxflow routing algorithm for wsns: practical issues and simulation-based assessment, *Proceedings of the 5th international conference on Soft computing as transdisciplinary science and technology*, pp. 688–693.

Shafiullah, G. M., Gyasi-Agyei, A. & Wolfs, P. J. (2008). *A Survey of Energy-Efficient and QoS-Aware Routing Protocols for Wireless Sensor Networks*, Springer Netherlands.

Sudevalayam, S. & Kulkarni, P. (2010). Energy harvesting sensor nodes: Survey and implications, *Communications Surveys and Tutorials, IEEE* pp. 1–19.

Wang, A. Y. & Sodini, C. G. (2006). On the energy efficiency of wireless transceivers, *Proc. of IEEE Conference on Communications*, pp. 3783–3788.

Yarvis, M. & Zorzi, M. (2008). Special issue on energy efficient design in wireless ad hoc and sensor networks, *Elsevier Ad Hoc Networks* .

Yick, J., Mukherjee, B. & Ghosal, D. (2008). Wireless sensor network survey, *Elsevier Computer Networks* 52: 2292–2330.

Zeng, K., Ren, K., Lou, W. & Moran, P. J. (2006). Energy-aware geographic routing in lossy wireless sensor networks with environmental energy supply, *Proceedings of the 3rd international conference on Quality of service in heterogeneous wired/wireless networks*.

8

Vibration Energy Harvesting:
Linear and Nonlinear Oscillator Approaches

Luca Gammaitoni, Helios Vocca, Igor Neri,
Flavio Travasso and Francesco Orfei
NiPS Laboratory – Dipartimento di Fisica, Università di Perugia,
INFN Perugia and Wisepower srl
Italy

1. Introduction

An important question that must be addressed by any energy harvesting technology is related to the type of energy available (Paradiso et al., 2005; Roundy et al., 2003). Among the renewable energy sources, kinetic energy is undoubtedly the most widely studied for applications to the micro-energy generation[1]. Kinetic energy harvesting requires a transduction mechanism to generate electrical energy from motion. This can happen via a mechanical coupling between the moving body and a physical device that is capable of generating electricity in the form of an electric current or of a voltage difference. In other words a kinetic energy harvester consists of a mechanical moving device that converts displacement into electric charge separation.

The design of the mechanical device is accomplished with the aim of maximising the coupling between the kinetic energy source and the transduction mechanism.

In general the transduction mechanism can generate electricity by exploiting the motion induced by the vibration source into the mechanical system coupled to it. This motion induces displacement of mechanical components and it is customary to exploit relative displacements, velocities or strains in these components.

Relative displacements are usually exploited when electrostatic transduction is considered. In this case two or more electrically charged components move performing work against the electrical forces. This work can be harvested in the form of a varying potential at the terminals of a capacitor.

Velocities are better exploited when electromagnetic induction is the transduction mechanism under consideration. In this case the variation of the magnetic flux through a coil due to the motion of a permanent magnet nearby is exploited in the form of an electric current through the coil itself.

---

[1] Clearly kinetic energy is not the only form of energy available at micro and nanoscale. As an example light is a potentially interesting source of energy and nanowires have been studied also in this respect (see e.g. Tian, B. Z. et al. *Coaxial silicon nanowires as solar cells and nanoelectronic power sources*. Nature 449, 885–890, 2007), however in this chapter we will focus on kinetic energy only.

Finally, strains are considered when the transduction mechanism is based on piezoelectric effects. Here electric polarization proportional to the strain appears at the boundaries of a strained piezo material. Such a polarization can be exploited in the form of a voltage at the terminals of an electric load.

In this chapter we will be mainly dealing with transduction mechanisms based on the exploitation of strains although most of the conclusions obtained conserve their validity if applied also to the other two mechanisms.

Before focusing on the characters of the available vibrations it is worth considering the energy balance we are facing in the energy harvesting problem. In the following figure we summarize the balance of the energies involved.

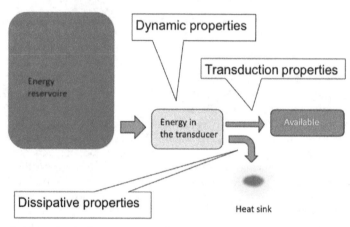

Fig. 1. Energy balance for the harvesting problem.

The kinetic energy available in the environment (red tank) inputs the transducer under the form of work done by the vibrational force to displace the mechanical components of the harvester. At the equilibrium, part of this energy is stored in the device as elastic and/or kinetic energy, part is dissipated in the form of heat and finally part of it is transduced into electric energy available for further use. Different fractions of the incoming energy have different destinations. The relative amount of the different parts depends on the dynamic properties of the transducer, its dissipative and transduction properties, each of them playing a specific role in the energy transformation process. We will come back to the energy balance problem below, when we will deal with a specific transduction mechanism.

2. The character of available energies

At micro and nanoscale kinetic energy is usually available as random vibrations or displacement noise. It is well known that vibrations potentially suitable for energy harvesting can be found in numerous aspects of human experience, including natural events (seismic motion, wind, water tides), common household goods (fridges, fans, washing machines, microwave ovens etc), industrial plant equipment, moving structures such as automobiles and aeroplanes and structures such as buildings and bridges. Also human and animal bodies are considered interesting sites for vibration harvesting. As an example in Fig. 2 we present three different spectra computed from vibrations taken from a car hood in motion, an operating microwave oven and a running train floor.

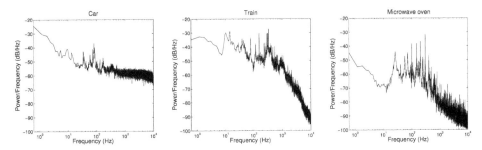

Fig. 2. Vibration power spectra. Figure shows acceleration magnitudes (in db/Hz) vs frequency for three different environments.

All these different sources produce vibrations that vary largely in amplitude and spectral characteristics. Generally speaking the human motion is classified among the high-amplitude/low-frequency sources. These very distinct behaviours in the vibration energy sources available in the environment reflect the difficulty of providing a general viable solution to the problem of vibration energy harvesting.

Indeed one of the main difficulties that faces the layman that wants to build a working vibration harvester is the choice of a suitable vibration to be used as a test bench for testing his/hers own device. In the literature is very common to consider a very special vibration signal represented by a sinusoidal signal of a given frequency and amplitude. As in (Roundy et al. 2004) where a vibration source of 2.5m s−2 at 120Hz has been chosen as a baseline with which to compare generators of differing designs and technologies. Although this is a well known signal that being deterministic in its character (can provide an easy approach both for generation and also for mathematical treatment), the results obtained with this signal are very seldom useful when applied to operative conditions where the vibration signal comes in the form of a random vibration with broad and often non stationary spectra. As we will see more extensively below, the specific features of the vibration spectrum can play a major role in determining the effectiveness of the harvester. In fact, most of the harvesters presently available in the market are based on resonant oscillators whose oscillating amplitude can be significantly enhanced due to vibrations present at the oscillator resonance frequency. On the other hand this kind of harvester results to be almost insensitive to vibrations that fall outside the usually narrow band of its resonance. For this reason it is highly recommended that the oscillator is built with specific care at tuning the resonance frequency in a region where the vibration spectrum is especially rich in energy. As a consequence it is clear that a detailed knowledge of the spectral properties of the hunted vibrations is of paramount importance for the success of the harvester. Unfortunately there is only a limited amount of public knowledge available of the spectral properties of vibrations widely available. In order to fill this gap, a database that collects time series from a wide variety of vibrating bodies was created (Neri et al. 2011). The database, still in the "accumulation phase" has to be remotely accessible without dedicated software. For this reason a data presentation via a web interface is implemented.

3. Micro energies for micro devices and below

An interesting limiting case of kinetic energy present in the form of random vibration is represented by the thermal fluctuations at the nanoscale. This very special environment represents also an important link between the two most promising sources of energy at the nanoscale: thermal gradients and thermal non-equilibrium fluctuations (Casati 2008).

Energy management issues at nanoscales require a careful approach. At the nanoscale, in fact, thermal fluctuations dominates the dynamics and concepts like "energy efficiency" and work-heat relations imply new assumptions and new interpretations. In recent years, assisted by new research tools (Ritort 2005), scientists have begun to study nanoscale interactions in detail. Non-equilibrium work relations, mainly in the form of "fluctuation theorems", have shown to provide valuable information on the role of non-equilibrium fluctuations. This new branch of the fluctuation theory was formalized in the chaotic hypothesis by Gallavotti and Cohen (Gallavotti 1995). Independently, Jarzynski and, then, Crooks derived interesting equalities (Jarzynski 1997), which hold for both closed and open classical statistical systems: such equalities relate the difference of two equilibrium free energies to the expectation of an ad hoc stylized non-equilibrium work functional.

In order to explore viable solutions to the harvesting of energies down to the nanoscales a number of different routes are currently explored by researchers worldwide. An interesting approach has been recently proposed within the framework of the race "Toward Zero-Power ICT"[2]. Within this perspective three classes of devices have been recently proposed (Gammaitoni et al., 2010):

- Phonon rectifiers
- Quantum harvesters
- Nanomechanical nonlinear vibration oscillators

The first device class (**Phonon rectifiers**) deals with the exploitation of thermal gradients (here interpreted in terms of phonon dynamics) via spatial or time asymmetries. The possibility of extracting useful work out of unbiased random fluctuations (often called *noise rectification*) by means of a device where all applied forces and temperature gradients average out to zero, can be considered an educated guess, for a rigorous proof can hardly be given. P. Curie postulated that if such a venue is not explicitly ruled out by the dynamical symmetries of the underlying microscopic processes, then it will generically occur.

The most obvious asymmetry one can try to advocate is spatial asymmetry (say, under mirror reflection, or chiral). Yet, despite the broken spatial symmetry, equilibrium fluctuations alone cannot power a device in a preferential direction of motion, lest it operates as a Maxwell demon, or *perpetuum mobile* of the second kind, in apparent conflict with the Second Law of thermodynamics. This objection, however, can be reconciled with Curie's criterion: indeed, a necessary (but not sufficient!) condition for a system to be at thermal equilibrium can also be expressed in the form of a dynamical symmetry, namely reversibility, or time inversion symmetry (detailed balance). Time asymmetry is thus a second crucial ingredient one advocates in the quest for noise rectification. Note, however,

---

[2] "Toward Zero-Power ICT" is an initiative of Future and Emerging Technologies (FET) program within the ICT theme of the Seventh Framework Program for Research of the European Commission.

that detailed balance is a subtle probabilistic concept, which, in certain situations, is at odds with one's intuition. For instance, as reversibility is not a sufficient equilibrium condition, rectification may be suppressed also in the presence of time asymmetry. On the other hand, a device surely operates under non-equilibrium conditions when stationary external perturbations act directly on it or on its environment.

In the concept device studied by the NANOPOWER project the asymmetry is related to the discreteness of phonon modes in cavities. By playing on the mismatch between the energy levels between a small cavity and a bigger one enabling the continuum to be reached, one could find that the transmission are not equal from left to right and vice versa. Phonon rectification occurs as the heat flow between the cavities becomes imbalanced due to non-matching phonon energy levels in it.

Quantum harvesters is a novel class of devices based on mesoscopic systems where unconventional quantum effects dominate the device dynamics. A significant example of this new device class is the so-called Buttiker-Landauer motor (Benjamin 2008) based on a working principle proposed by M. Buttiker (Buttiker 1987) and dealing with a Brownian particle moving in a sinusoidal potential and subject to non-equilibrium noise and a periodic potential. The motion of an underdamped classical particle subject to such a periodic environmental temperature modulation was investigated by Blanter and Buttiker (Blanter 1998). Recently this phenomenology has been experimentally investigated in a system of electrons moving in a semiconductor system with periodic grating and subjected terahertz radiation (Olbrich et al. 2009). The grating is shaped in such a way that it provides both the spatial variation for electron motion as well as a means to absorb radiation of much longer wavelength than the period of the grating.

The third device class is represented by **nano-mechanical nonlinear vibration oscillators**. Nanoscale oscillators have been considered a promising solution for the harvesting of small random vibrations of the kind described above since few years. A significant contribution to this area has been given by Zhong Lin Wang and colleagues at the Georgia Institute of Technology (Wang et al. 2006). In a recent work (Xu et al. 2010) they grew vertical lead zirconate titanate (PZT) nanowires and, exploiting piezoelectric properties of layered arrays of these structures, showed that can convert mechanical strain into electrical energy capable of powering a commercial diode intermittently with operation power up to 6 mW. The typical diameter of the nanowires is 30 to 100 nm, and they measure 1 to 3 μm in length.

A different nano-mechanical generator has been realized by Xi Chen and co-workers (Chen Xi et al. 2010), based on PZT nanofibers, with a diameter and length of approximately 60 nm and >500 μm, aligned on Platinum interdigitated electrodes and packaged in a soft polymer on a silicon substrate. The PZT nanofibers employed in this generator have been prepared by electrospinning process and exhibit an extremely high piezoelectric voltage constant (g33: 0.079 Vm/N) with high bending flexibility and high mechanical strength (unlike bulk, thin films or microfibers). Also Zinc-Oxide (ZnO) material received significant attention in the attempt to realize reliable nano-generators. Min-Yeol Choi and co-workers (Min-Yeol Choi et al. 2009) have recently proposed a transparent and flexible piezoelectric generator based on ZnO nanorods. The nanorods are vertically sandwiched between two flat surfaces producing a thin mattress-like structure. When the structure is bended the nanorods get compressed and a voltage appear at their ends.

At difference with these existing approaches, in the NANOPOWER project attention is focussed mainly on the dynamics of nanoscale structures and for a reason that will be discussed below, it concentrates in geometries that allowed a clear nonlinear dynamical behaviour, like bistable membranes. Recently (Cottone et al. 2009, Gammaitoni et al. 2009) a general class of bistable/multistable nonlinear oscillators have been demonstrated to have noise-activated switching with an increased energy conversion efficiency. In order to reach multi-stable operation condition, in NANOPOWER a clamped membrane is realised under a small compressive strain, forcing it to either of the two positions. The membrane vibrates between the two potential minima and has also intra-minima modes. The kinetic energy of the nonlinear vibration is converted into electric energy by piezo membrane sandwiched between the electrodes.

4. Fundaments of vibration harvesting

As we have anticipated above, kinetic energy harvesting requires a transduction mechanism to generate electrical energy from motion. This is typically achieved by means of a transduction mechanism consisting in a massive mechanical component attached to an inertial frame that acts as the fixed reference. The inertial frame transmits the vibrations to a suspended mass, producing a relative displacement between them.

4.1 A simple scheme for vibration harvesting

The scheme reproduced in Fig. 3 shows the inertial mass m that is acted on by the vibrations transmitted by the vibrating body to the reference frame.

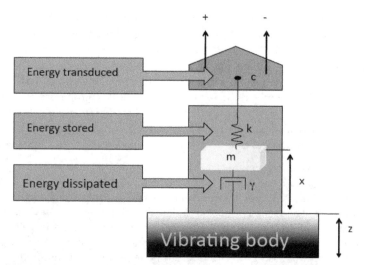

Fig. 3. Vibration-to-electricity dynamic conversion scheme. Energy balance: the kinetic energy input into the system from the contact with the vibrating body is partially stored into the system dynamics (potential energy of the spring), partially dissipated through the dashpot and partially transduced into electric energy available for powering electronic devices.

In terms of energy balance, the input energy, represented by the kinetic energy of the vibrating body, is transmitted to the harvester. This input energy is divided into three main components:

1. Part of the energy is **stored** into the dynamics of the mass and is usually expressed as the sum of its kinetic and potential energy: when the spring is completely extended (or compressed), the mass is at rest and all the dynamic energy is represented by the potential (elastic) energy of the spring.

2. Part of the energy is **dissipated** during the dynamics meaning with this that it is converted from kinetic energy of a macroscopic degree of freedom into heat, i.e. the kinetic energy of many microscopic degrees of freedom. This is represented in Fig. 3 by the dashpot. There are different kinds of dissipative effects that can be relevant for a vibration harvester. One simple example is the internal friction of the material undergoing flexure. Another common case is *viscous damping* a source of dissipation due to the fact that the mass is moving within a gas and the gas opposes some resistance.

3. Finally, some of the energy is **transduced** into electric energy. The transducer is represented in Fig. 3 by the block with the two terminals + and -, thus indicating the existence of a voltage difference V.

4.2 A mathematical model for our scheme

The functioning of the vibration harvester, within this scheme, can usually be quantitatively described in terms of a simple mathematical model that addresses the dynamics of the two relevant quantities: the mass displacement x and the voltage difference V. Both quantities are function of time and obey proper equations of motion.

For the displacement x the dynamics is described by the standard Newton equation, i.e. a second order ordinary differential equation:

$$m\ddot{x} = -\frac{d}{dx}U(x) - \gamma\dot{x} - c(x,V) + \xi_z \qquad (1)$$

where

$U(x)$	Represents the energy stored
$\gamma\dot{x}$	Represents the dissipative force
$c(x,V)$	Represents the reaction force due to the transduction mechanism
ξ_z	Represents the vibration force

The quantity ξ_z represents here the vibration force that acts on the oscillator. In general this is a stochastic quantity due to the random character of practically available vibrations. For this reason the equation of motion cannot be considered an ordinary differential equation but it is more suitably defined a stochastic differential equation, also know as *Langevin equation* by the name of the French physicist who introduced it in 1908 in order to describe the Brownian motion (Langevin, 1908).

All the components of the energy budget that we mentioned above are in this equation represented in term of forces. In particular the quantity γ is the damping constant and multiplies the time derivative of the displacement, i.e. the velocity. Thus this term represents a dissipative force that opposes the motion with an intensity proportional to the velocity: a condition typical of viscous damping (the faster the motion the greater the force that opposes it).

The quantity $c(x,V)$ is a general function that represents the reaction force due to the motion-to-electricity conversion mechanism. It has the same sign of the dissipative force and thus opposes the motion. In physical terms this arises from the energy fraction that is taken from the kinetic energy and transduced into electric energy.

The dynamics of the voltage V is described by:

$$\dot{V} = F(\dot{x},V) \tag{2}$$

This is a first order differential equation that connects the velocity of the displacement with the electric voltage generated. In order to reach a full description of the motion-to-electric-energy conversion we need to specify the form of the two connecting functions

$$F(\dot{x},V), \quad c(x,V)$$

These two functions are determined once we specify the physical mechanism employed to transform kinetic energy into electric energy.

4.3 The piezoelectric transduction case

As we pointed out earlier there are three main physical mechanisms that are usually considered at this aim: piezoelectric conversion (dynamical strain of piezo material is converted into voltage difference), electromagnetic induction (motion of magnets induces electric current in coils) and capacitive coupling (geometrical variations of capacitors induce voltage difference).

For a number of practical reasons (Roundy et al. 2003) mainly related to the possibility to miniaturize the generator maintaining an efficient energy conversion process, piezoelectricity is generally considered the most interesting mechanism.

For the case of piezoelectric conversion the two connecting functions assume a simple expression:

$$c(x,V) = K_V V$$
$$F(\dot{x},V) = K_c \dot{x} - \frac{1}{\tau_p} V$$

The dynamical equations thus become:

$$m\ddot{x} = -\frac{d}{dx}U(x) - \gamma\dot{x} - K_V V + \xi_z$$
$$\dot{V} = K_c \dot{x} - \frac{1}{\tau_p} V \tag{3}$$

where K_V and K_c are piezoelectric parameters that depend on the physical properties of the piezo material employed and τ_p is a time constant that can be expressed in terms of the parameters of the electric circuit connected to the generator as:

$$\tau_p = R_L C \tag{4}$$

where C is the electrical capacitance of the piezo component and R_L is the load resistance across which the voltage V is exploited. In this scheme the power extracted from the harvester is given by

$$W = \frac{V^2}{R_L} \tag{5}$$

5. The linear oscillator approach: Performances and limitations

In order to proceed further in our analysis of the vibration harvester we need to focus our attention on the quantity $U(x)$ that following the schematic in Fig. 3, represents the potential energy. The mathematical form of this function is a consequence of the geometry and of the dynamics of the vibration harvester that we want to address.

One of the most common models of harvester is the so-called *cantilever configuration*. A typical cantilever is reproduced in Fig. 4.

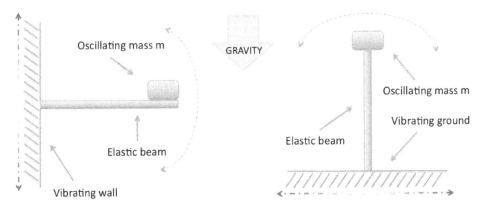

Fig. 4. Vibration energy harvester represented here as a cantilever system. Left: configuration for harvesting vertical vibrations. Right: configuration for harvesting horizontal vibrations.

According to our schematic in Fig. 3, the "spring like" behaviour of the harvester is represented here by the bending of the beam composing the cantilever. When the beam is completely bent (corresponding to the case where the spring is completely extended or compressed), as we have seen above, the mass is at rest and all the dynamic energy is represented by the (potential) elastic energy.

A common assumption is that the potential energy grows with the square of the bending. This is based on the idea that the force acted by the beam is proportional to the bending.

Thus is $F = kx$ than $U(x) = -1/2\ k\ x^2$. The idea that the force is proportional to the bending is quite reasonable and has been verified in a number of different cases. An historically relevant example is the Galileo's pendulum. In this case the "bending" is represented by the displacement of the mass from the vertical position. Due to the action of the gravity, the restoring force that acts on the pendulum mass is $F = -\ mg\ sin(x/l)$ where g is the gravity acceleration and l is the pendulum length. When the displacement angle x/l is small the function $sin(x/l)$ is approximately equal to x/l (first term of the Taylor expansion around $x/l = 0$) and thus $F = -\ mg/l\ x$ or $F = -\ k\ x$. This condition is usually known as "small oscillation approximation" and can be applied any time we have a *small*[3] oscillation condition around an equilibrium point.

5.1 The linear vibration harvester

Within the small oscillation approximation we can treat most of the vibration energy harvesters by introducing a potential energy function like the following:

$$U(x) = \frac{1}{2}kx^2 \qquad (6)$$

This form is also known as *harmonic potential*. By substituting (6) in (3) and taking the derivative, the equations of motion now become:

$$m\ddot{x} = -kx - \gamma\dot{x} - K_V V + \xi_z$$
$$\dot{V} = K_c\dot{x} - \frac{1}{\tau_p}V \qquad (7)$$

This often called the linear oscillator approximation, due to the fact that, as we have seen, in the dynamical equation of the displacement x the force is linearly proportional to the displacement itself.

A linear oscillator is a very well known case of Newton dynamics and its solution is usually studied in first year course in Physics. A remarkable feature of a linear oscillator is the existence of the *resonance frequency*. When the system is driven by a periodic external force with frequency equal to the resonance frequency then the system response reaches the maximum amplitude.

This occurrence is well described by the so-called system transfer function $H(w)$ whose study is part of the *linear response theory* addressed in physics and engineering course in dynamical systems.

A detailed treatment of the linear response theory is well beyond the scope of this chapter. For our purposes it is sufficient to observe that a linear system represents a good approximation of a number of real oscillators (in the small oscillation approximation) and that their behaviour is characterized by the existence of a resonance frequency that maximizes the oscillator amplitude. This condition has led vibration energy harvester designer to try to build cantilevers (Williams CB et al., 1996, Mitcheson et al., 2004, Stephen

---

[3] The oscillation angle is considered small when the terms following the first one in the Taylor expansion of the sine are negligible compare to the leading one

N.G., 2006, Renno J.M. et al., 2009) that operates in the linear oscillation regime and present a resonant frequency that can be tuned to match the characteristic frequency of the application environment. Various attempts to design tuneable harvesters have been proposed (Challa V.R. et al., 2008, Morris D. et al., 2008). Thus most of the present working solutions for vibration-to-electricity conversion are based on linear, i.e. resonant, mechanical oscillators that convert kinetic energy by tuning their resonant frequency in the spectral region where most of the energy is available. However, as we have observed above, in the vast majority of cases the ambient vibrations have their energy distributed over a wide frequency spectrum, with significant predominance of low frequency components and frequency tuning is not always possible due to geometrical/dynamical constraints.

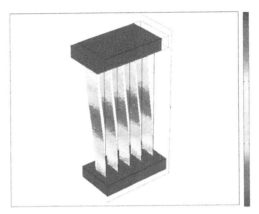

Fig. 5. Composite model of a microscale linear vibration energy harvester. The five beams represented here are 1 mm high and 25 µm thick. The resonant frequency (fundamental mode) for this system is as high as 10 KHz. Colours represents the map of stresses during the dynamics.

5.2 Main limitations in the linear vibration harvester

In general, for a generic linear system, even more complicated that the simple cantilever in Fig. 4, the transfer function presents one or more peaks corresponding to the resonance frequencies and thus it is effective mainly when the incoming energy is abundant in that frequency regions.

Unfortunately this is a serious limitation when it is required to build a vibration energy harvester of small dimension, for at least two main reasons:

a. as we have discussed above, the frequency spectrum of ambient available vibrations instead of being sharply peaked at some frequency is usually very broad.
b. The frequency spectrum of available vibrations is particularly rich in energy in the low frequency range, and it is very difficult, if not impossible, to build small, low-frequency resonant systems.

As an example we can consider the composite system sketched in Fig. 5. In this case the resonant frequency (fundamental mode) is of the order of 10 KHz, a frequency region where in most practical cases the presence of ambient vibrations is almost negligible. As a consequence we are forced to build vibration harvesters that have geometrical dimensions compatible with low resonance frequencies. Also in this case, however there is a serious limitation arising from the need to tune the resonance frequency according to the specific application. As an example we can consider the following situation: we have designed a vibration harvester that is supposed to operate at 100 Hz (a typical value for commercial harvesters of few centimetres) because for the specific application that we are considering there is enough vibration amplitude at that very frequency. If the ambient condition changes, e.g. the frequency peak of the ambient vibration moves from 100 Hz to 80 Hz our harvester needs a tuning operation in order to lower its resonant frequency. Such an operation can be made, not without difficulties, by changing some physical parameters, like the length or the stiffness of the cantilever beam or the mass attached to its tip. These operations are clearly not easy to perform and thus a linear vibration harvester in general does not allow for wide *tunability*.

5.3 In search of the ideal vibration harvester

Based on the considerations developed so far it is clear that vibration harvesters inspired by cantilever-like configurations present a number of drawbacks that limit seriously their field of application. If we try to summarize what we have learned so far we see that the search for the ideal vibration harvester have to cover at least the following aspects:

- Capability of harvesting energy on a broadband of frequencies and not just at the resonance frequency. This seems to exclude resonant oscillators. In fact a resonant oscillator is capable of harvesting energy only in a very narrow band, i.e. around its resonant frequency. Moreover for a linear oscillator the narrower is the resonance the better is the efficiency in harvesting energy at that frequency. Thus, if we want to keep the requirement 1) we should look for non-resonant oscillators.
- Capability of harvesting energy at low frequency (below few hundred Hz) because this is where most of the ambient available energy is. Also in this case, due to the conflicting requirements of small dimensions and low operating frequencies, we should abandon linear, i.e. resonant oscillators for some other dynamical system that shows a significant response also at low frequencies.
- No need for frequency tuning after the initial set-up of the harvester. Frequency tuning is a feature of resonant systems, thus if we move to non-resonant systems this requirement will be automatically satisfied.

As we have seen before the search for the best solution in terms of non-resonant systems should start from the potential energy function $U(x)$. In fact $U(x)$ plays the role of dynamical energy storage facility (before transduction) for our mechanical oscillator and thus it is here that we should focus our attention.

6. The nonlinear oscillator approach: Performances and limitations

The natural candidates to replace linear oscillators seem clearly to be the so called "non-linear" oscillators. Unfortunately this is not a well-defined category and one is left with the sole condition

$$U(x) \neq \frac{1}{2}kx^2 \qquad (8)$$

meaning with this expression oscillators whose potential energy is not quadratically dependent on the relevant displacement variable. In recent years few possible candidates have been explored (Cottone et al., 2009; Gammaitoni et al., 2009, 2010, Ferrari M. et al, 2009, Arrieta A.F. et al., 2010, Ando B. et al., 2010, Barton D.A.W. et al., 2010, Stanton S.C. et al., 2010) running from

$$U(x) = ax^{2n} \qquad (9)$$

to other more complicated expressions.

For nonlinear oscillators it is not possible to define a transfer function (i.e. a response function independent from the external force acting on the system) and thus a properly defined resonant frequency even if the power spectrum density of the system can show one or more well defined peaks for specific values of the frequencies.

In this section we will briefly address one of these nonlinear potential cases, with the specific purpose of describing and testing the power generated in common environments.

Specifically we will show that, if we consider bistable or monostable oscillators under proper operating conditions, they can provide better performances compared to a linear oscillator in terms of energy extracted from a generic wide spectrum vibration.

6.1 The bistable cantilever

An interesting option for a nonlinear oscillator is to look for a potential energy that is multi-stable, instead of mono-stable (like the linear case, i.e. the harmonic potential). A particularly simple and instructive example on how to move from the linear (mono-stable) to a possible nonlinear (bi-stable) dynamics is represented by a slightly modified version of our vibration harvester cantilever (see Fig. 6).

This is a common cantilever operated vertically (in a configuration sometimes called *inverted pendulum configuration*) with a bending piezoelectric beam. On top of the beam mass a small magnet (tip magnet) has been added. Under the action of the vibrating ground the pendulum oscillates alternatively bending the piezoelectric beam and thus generating a measurable voltage signal V. The dynamics of the inverted pendulum tip can be controlled with the introduction of an external magnet conveniently placed at a certain distance D and with polarities opposed to those of the tip magnet. The external magnet introduces a force dependent from the distance that opposes the elastic restoring force of the bended beam. As a result, the inverted pendulum dynamics can show three different types of behaviours as a function of the distance D:

- when the two magnets are separated by a large distance ($D \gg D_0$) the inverted pendulum behaves like a linear oscillator whose dynamics is resonant with a resonance frequency determined by the system parameters. This situation accounts well for the usual operating condition of traditional piezoelectric vibration-to-electric energy converters.
- On the other hand, when D is small ($D \ll D_0$) the pendulum is forced to oscillates at the left or at the right of the vertical. In the limit of small oscillations, this can still be described in terms of a linear, i.e. resonant, oscillator with a resonant frequency higher that in the previous case.

- Finally, in between the two previous cases it exists an intermediate condition ($D=D_0$) where the pendulum swings in a more complex way with small oscillations around each of the two equilibrium positions (left and right of the vertical) and large excursions from one to the other.

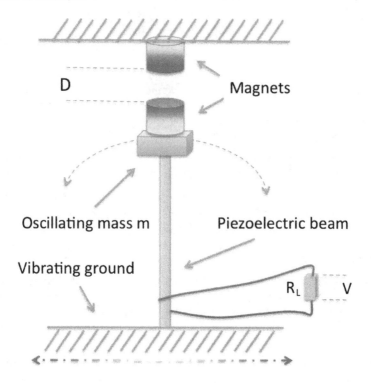

Fig. 6. Piezoelectric inverted pendulum showing bistable dynamics.

In Fig. 7, 8 and 9 we present an analysis of the simulated dynamics with a modified potential that takes into account the presence of the repulsion force due to the magnets. This new potential can be written as:

$$U(x) = \frac{1}{2}k_e x^2 + (Ax^2 + BD^2)^{-\frac{3}{2}} \qquad (10)$$

with k_e, A and B representing constants related to the physical parameters of the pendulum (Lefeuvre et al., 2006, Shu et al., 2006) like the permeability constant and the effective magnetic moment of the magnets. Clearly when the distance D between the magnets grows very large the second term in (10) becomes negligible and the potential tends to the harmonic potential of the linear case, typical of the cantilever harvester.

This is the condition represented in Fig. 7 (top) where we plot the potential $U(x)$ vs x. Under the picture of the potential we plot the displacement x time series. This is measured in meters and it is referred to an inverted pendulum of few centimetres long. For details please see (Cottone et al. 2009). Further below there is the time series for the voltage V across the load resistor R_L.

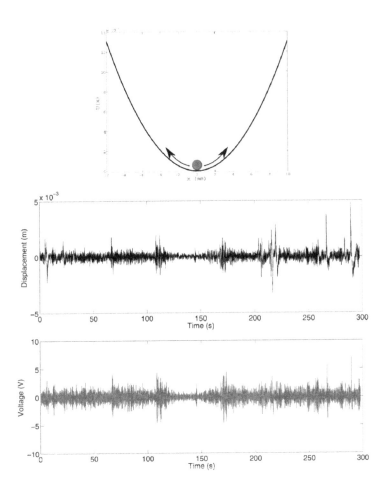

Fig. 7. Upper panel: potential energy $U(x)$ in (10) in arbitrary units when (D>>D_0). Middle panel: displacement x time series and Lower panel: voltage V time series. Both quantities have been obtained via a numerical solution of the stochastic differential equation (3) with potential (10). The stochastic force is an exponentially correlated noise with fixed standard deviation and correlation time 0.1 s.

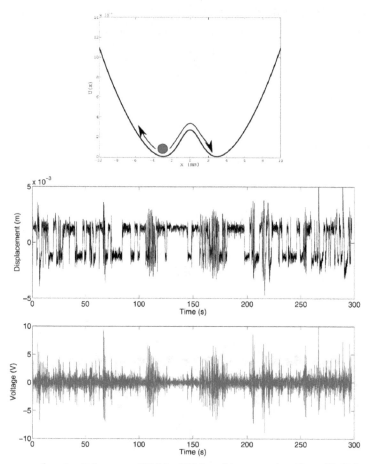

Fig. 8. Upper panel: potential energy $U(x)$ in (10) in arbitrary units when $(D=D_0)$.
Middle panel: displacement x time series and Lower panel: voltage V time series. Both
quantities have been obtained via a numerical solution of the stochastic differential equation
(3) with potential (10). The stochastic force is an exponentially correlated noise with fixed
standard deviation and correlation time 0.1 s.

It is worth notice that the voltage V, obtained from the solution of the coupled differential
stochastic equations (3), follows quite closely the displacement x time series. This is due to
the special form of the voltage dynamics equation:

$$\dot{V} = K_c \dot{x} - \frac{1}{\tau_p} V \tag{11}$$

In fact such an equation represent an high-pass filter whose input signal is the displacement
derivative (the velocity) and where the cut-on frequency is represented by $1/\tau_p$ with τ_p
given by (4). By Physical point of view this means that the piezoelectric material introduces

a characteristic time constant determined among other things by the physical properties of the piezoelectricity mechanism, i.e. the rearrangement dynamics of the internal electric dipole moments, once the strain is applied. In other words if we produce a bending of the beam too slowly the charges accumulated are easily dispersed and little or no global voltage difference V appears at the ends of the sample. On the other hand, if the time constant τ_p is very large compared to the strain dynamics, then the second term in (11) becomes negligible and V is simply proportional to x.

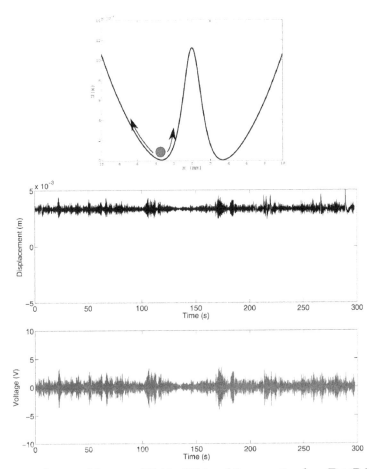

Fig. 9. Upper panel: potential energy $U(x)$ in (10) in arbitrary units when (D<<D_0). Middle panel: displacement x time series and Lower panel: voltage V time series. Both quantities have been obtained via a numerical solution of the stochastic differential equation (3) with potential (10). The stochastic force is an exponentially correlated noise with fixed standard deviation and correlation time 0.1 s.

In Fig. 8 we present the same quantities of Fig. 7, with the only difference that now the two magnets are not far away from each other but at a certain distance $D=D_0$. In this case the potential energy shows clearly two distinct equilibrium points separated by an energy barrier. The displacement dynamics (below) shows an evident bistable character with the displacement x that switches frequently between the two corresponding potential minima positions. As a consequence the voltage V (further below) shows a random behavior with large excursions in correspondence with the switches.

Finally, in Fig. 9, on further decreasing the distance of the two magnets (D << D_0) the potential energy becomes even more bistable and the barrier grows up to a point in which the jumps between the two minima become fewer and less probable. The displacement dynamics gets confined in one well and it shows lower amplitude that reflects into a smaller voltage amplitude V.

This overall qualitative behaviour is somehow summarized in a more quantitative way in Fig. 10, where the average power (average V^2/R_L) extracted from this vibration harvester is presented as a function of the distance parameter D. As it is well evident there is an optimal distance D_0 where the power peaks to a maximum. Most importantly such a maximum condition is reached in a full nonlinear regime (bistable condition of the potential) and results to be quite larger (at least a factor 4) than the value in the linear operation condition (far right in Fig. 10).

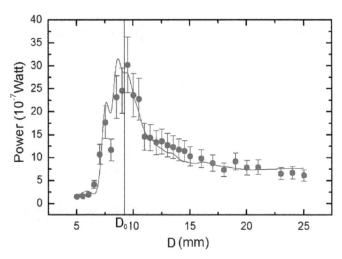

Fig. 10. Piezoelectric nonlinear vibration harvester mean electric power as a function of the distance D between the two magnets. The symbols correspond to experimental values measured from a prototype apparatus (see Cottone et al., 2009). The continuous curve has been obtained from the numerical solution of the stochastic differential equation (3) with potential (10). Both in the experiment and in the numerical solution, the stochastic force has the same statistical properties: an exponentially correlated signal with correlation time 0.1 s. Every data point is obtained from averaging the rms values of ten time series sampled at a frequency of 1 kHz for 200 s. The rms is computed after zero averaging the time series. The expected relative error in the numerical solution is within 10%. For further details on the numerical quantities please see (Cottone et al. 2009).

6.2 More general potentials

The results presented here show that, following our previous reasoning, there might be nonlinear potentials that allow better performances in terms of power generated by a vibration energy harvester, compared to the standard linear cantilever. Although the shape of potential in (10) is quite peculiar, there are more general potentials that allow good performances as well. We have explored few simpler forms, both in the bistable and in the monostable configuration. Among the bistable ones, a popular form is represented by the well know "Duffing oscillator" described by:

$$U(x) = \frac{1}{2}ax^2 + \frac{1}{4}bx^4 \tag{12}$$

This potential $U(x)$ is a good approximation to a number of practically available nonlinear oscillators, like pre-bended membranes or beams.

Among the monostable (nonlinear) oscillators, as previously anticipated, we have studied the simple class of potentials in (9) with a > 0 and n = 1,2,.. The n = 1 case corresponds to the harmonic potential. In this case the dynamics is the standard linear oscillator dynamics. For n > 1 we still have a monostable potential but the dynamics is not linear anymore. In Fig. 11 we show the displacement (x_{rms}) as a function of a and n. It is evident that if a is larger than a certain threshold (a_{th}) then x_{rms} increases with n and a nonlinear potential can easily outperform a linear one, thus extending the main finding of the bistable potential (10) also to the monostable case. On the other hand if $a < a_{th}$ then x_{rms} decreases with n and the linear case performs better than the nonlinear one.

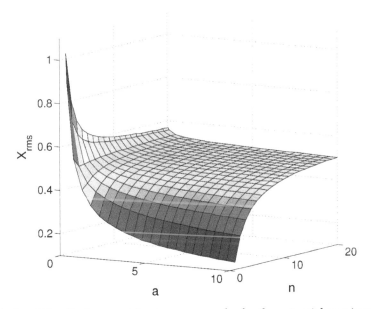

Fig. 11. 3D plot of the displacement (x_{rms}) versus a and n for the potential case in equation (9). For further details on numerical parameters please see (Gammaitoni et al. 2009, 2010).

It has been shown in (Gammaitoni et al. 2009, 2010) that the value a_{th} is linearly dependent on the ambient noise intensity. This result is relevant in the design of an efficient vibration harvesting device. In fact, in real-world applications, the value of the vibration intensity is set by the ambient and cannot be arbitrarily set. The value of the parameter a is usually fixed by the dynamical constraints or by the material properties (stiffness, inertia, etc.) while the value of n can sometimes be selected by a proper design of the geometry of the device itself. Thus, once the noise intensity and a are fixed the choice of a linear ($n = 1$) or nonlinear potential ($n > 1$) can be made in order to maximize the x_{rms} and consequently the power obtained at the device output.

7. Conclusion

In this chapter we have discussed problem of vibration energy harvesting with specific reference to the role of linearity and nonlinearity in the potential energy characterizing the mechanical transduction system. We have shown that linear cantilevers are clearly limited in their practical applications and that more complex nonlinear oscillators can outperform them in a number of realistic energy harvesting scenarios.

8. Acknowledgment

The results presented here have obtained thanks to the financial support from European Commission (FPVII, Grant agreement no: 256959, NANOPOWER and Grant agreement no: 270005, ZEROPOWER), Ministero Italiano della Ricerca Scientifica (PRIN 2007), and Fondazione Cassa di Risparmio di Perugia (Bando a tema - Ricerca di Base 2009, Microgeneratori di energia di nuova concezione per l'alimentazione di dispositivi elettronici mobili).

9. References

Ando B., Baglio S., Trigona C., Dumas N., Latorre L., Nouet P., 2010, "Nonlinear mechanism in MEMS devices for energy harvesting applications", *Journal of Micromechanics and Microengineering*, Vol 20, 125020.

Arrieta A.F., Hagedorn P., Erturk A. and Inman D.J., 2010, "A piezoelectric bistable plate for nonlinear broadband energy harvesting", *Applied Physics Letters*, Vol 97, 104102.

Barton D.A.W., Burrow S.G. and Clare L.R., 2010, "Energy Harvesting from Vibrations with a Nonlinear Oscillator," *Journal of Vibration and Acoustics*, 132, 021009.

Benjamin R. & Kawai R. (2008). Inertial effects in the Buttiker-Landauer motor and refrigerator at the overdamped limit, *Phys. Rev. E* 77, 051132.

Blanter Ya.M. & Büttiker M. (1998). Rectification of Fluctuations in an Underdamped Ratchet, Phys. Rev. Lett. 81, 4040.

Büttiker M. (1987). Transport as a Consequence of State-Dependent Diffusion, *Z. Phys. B* 68, 161.

Casati G.; Mejia-Monasterio C.; Prosen T.; *Phys Rev Lett*, 101, 016601 (2008).

Challa V.R., Prasad M.G., Shi Y. and Fisher F.T., 2008, "A vibration energy harvesting device with bidirectional resonance frequency tunability", *Smart Materials and Structures*, Vol 17, 015035.

Chen Xi; Shiyou Xu; Nan Yao; Yong Shi (2010) 1.6 V Nanogenerator for Mechanical Energy Harvesting Using PZT Nanofibers. *Nano Lett.*, 10 (6), pp 2133–2137.

Cottone, F.; Vocca, H.; Gammaitoni, L. (2009). Nonlinear energy harvesting, *Phys. Rev. Lett.* 102, 080601.

F. Ritort, (2003). Work fluctuations, transient violations of the second law and free-energy recovery methods: Perspectives in Theory and Experiments, *Poincare Sem.* 2 193.

Ferrari M., Ferrari V., Guizzetti M., Ando B., Baglio S., Trigona C., 2009, "Improved energy harvesting from wideband vibrations by nonlinear piezoelectric converters", *Sensors and Actuators A: Physical*, Vol 162, No 2, pp. 425-431.

Gallavotti G.; Cohen E.G.D. (1995). Dynamical ensembles in nonequilibrium statistical mechanics, *Phys Rev Lett,* 2694.

Gammaitoni L.; Marchesoni F.; Worshech L.; Ahopelto J.; Sotomayor-Torres C.; Buttiker M. (2010). NANOPOWER: Nanoscale energy management for powering ICT devices. EC FET Proactive funded project (Objective ICT-2009 8.6 – Call 5, GA no: 256959). See www.nanopwr.eu.

Gammaitoni, L.; Neri, I.; Vocca, H. (2009) Nonlinear oscillators for vibration energy harvesting, *Appl. Phys. Lett.*, 94, 164102.

Gammaitoni, L.; Neri, I.; Vocca, H. (2010) The benefits of noise and nonlinearity: Extracting energy from random vibrations, *Chemical Physics,* V. 375, p.435–438.

Jarzinsky C., (1997). Nonequilibrium equality for free energy dfferences. *Phys Rev Lett,* 2690.

Langevin, P. (1908). "On the Theory of Brownian Motion". *C. R. Acad. Sci. (Paris)* 146: 530–533.

Lefeuvre E.; et al. (2006), *Sens. Actuators* A, Phys. 126, 405.

Min-Yeol Choi, Dukhyun Choi, Mi-Jin Jin, Insoo Kim, Sang-Hyeob Kim, Jae-Young Choi1, Sang Yoon Lee, Jong Min Kim, Sang-Woo Kim, Mechanically Powered Transparent Flexible Charge-Generating Nanodevices with Piezoelectric ZnO Nanorods, *Advanced Materials*, 21, 2185–2189, 2009.

Mitcheson, P.D. et al., Architectures for Vibration-Driven Micropower Generators, *J. Microelectromechanical Systems*, vol. 13, no. 3, 2004, pp. 429–440.

Morris D. et al., 2008, "A resonant frequency tunable, extensional mode piezoelectric vibration harvesting mechanism", *Smart Materials and Structures*, Vol 17, 065021.

Neri, I.; Mincigrucci R.; Travasso F.; Vocca H.; Gammaitoni L.; A real vibration database for kinetic energy harvesting applications. *Journal of Intelligent Material Systems and Structures* (June 2011). Submitted.

P. Olbrich P.; Ivchenko E.L.; Ravash R.; Feil T.; Danilov S.D.; Allerdings J.; Weiss D.; Schuh D.; Wegscheider W.; Ganichev S.D.; (2009). Ratchet Effects Induced by Terahertz Radiation in Heterostructures with a Lateral Periodic Potential, *Phys. Rev. Lett.* 103, 090603.

Paradiso J.A. & Starner T. (2005). Energy Scavenging for Mobile and Wireless Electronics, *IEEE Pervasive Computing* 4, 18.

Renno J.M., Daqaq M.F., Inman D.J., 2009, "On the optimal energy harvesting from a vibration source", *Journal of Sound and Vibration* Vol 320, No 1-2, pp. 386-405.

Roundy S. & Wright P.K. (2004). A piezoelectric vibration based generator for wireless electronics. *Smart Mater. Struct.* 13 1131–1142.

Roundy S. et al., A study of low level vibrations as a power source for wireless sensor nodes, *Computer Communications* 26 (2003) 1131–1144.

Roundy, S.; Wright, P.K.; Rabaey J.M. (2003). Energy Scavenging for Wireless Sensor Networks, *Kluwer Academic Publishers* Boston.

Shu, Y. C. and Lien, I. C. (2006), *J. Micromech. Microeng.* 16, 2429 (2006)

Stanton S.C., McGehee C.C. and Mann B.P., 2010, "Nonlinear Dynamics for Broadband Energy Harvesting: Investigation of a Bistable Inertial Generator," *Physica D*, 239, pp. 640-653.

Stephen N.G., 2006, "On energy harvesting from ambient vibration", *Journal of Sound and Vibration*, Vol 293, pp. 409-425.

Wang Z.L. & Song J.H. (2006). Piezoelectric Nanogenerators Based on Zinc Oxide Nanowire Arrays, *Science*, 312, 242-246.

Williams CB, Yates RB, 1996, "Analysis of a Micro-Electric Generator for Microsystems", *Sensors and Actuators A: Physical*, Vol 52, pp. 8–11.

Xu S.; Hansen B.; Wang Z.L. (2010). Piezoelectric-nanowire-enabled power source for driving wireless microelectronics *Nature Communications* 1 93.

9

Hydrogen from Stormy Oceans

Helmut Tributsch

Retired from Free University Berlin, Institute for Physical and Theoretical Chemistry, and Helmholtz-Centre Berlin for Materials and Energy, Germany

1. Introduction

The recent nuclear accident of Fukushima, Japan, which has hit a highly industrialized and technically advanced country, as a consequence of a natural disaster, has significantly altered worldwide opinions on possible energy strategies for the future. For many decision makers nuclear energy was a feasible large scale technology for bridging the time until cheap sustainable energy would be available sometime in the future. The advanced technology of nuclear energy and the high energy density generated convinced them in spite of the fact, that no safe solution is still available for handling radioactive waste and that the availability of uranium is limited. Now the prospect of operating up to 20.000 nuclear reactors on earth to provide a reasonable contribution to the world`s huge energy consumption (estimated 45 TW, terawatt, by the end of this century) is frightening many people. Highly developed countries like Germany, Swizzerland and Italy have already voted to search for an energy future without nuclear energy. This, however, poses a significant challenge. The development of sustainable energy sources must proceed in a much more aggressive way. And there is the question, whether highly subsidized solar technologies like photovoltaics can be preferred choices on a shorter term in a massive effort towards clean energy. Industrialized countries may be able to afford such subsidized energy, but poor countries are not. And what the industrialized world mostly needs are sustainable fuels for transport and chemical industry. What would be the most efficient and the least costly path towards a sustainable energy economy?

The author has recently studied this question and came to the conclusion that industrial society should make a bio-mimetic, or bionic approach for solving its energy problems (Tributsch 2011). Living nature has not only adapted an originally hostile climate of our planet to favourable living conditions. It has also succeeded in supplying to living beings both abundant fuels and chemicals in an entirely sustainable way. The strategy adopted was to split water with solar light, but then to attach the liberated hydrogen to a carbon containing energy carrier, carbon dioxide. Starting from the resulting carbohydrates all needed fuels and chemicals could be synthesized in an entirely sustainable way.

It is true that at present we do not have the technology to split water directly with light using a suitable catalyst. But we could use other technological strategies such as electricity generated from wind and other sustainable sources to produce hydrogen from water. Only gasified biomass is presently available as a sustainable carbon containing molecular carrier and this resource is limited so that artificial techniques for biomass generation will be needed in the future (Tributsch, 2011). Using Fischer –Tropsch synthesis pathways, hydrogen may be combined with gasified biomass to yield any type of carbon containing energy carrier or chemical.

From such biomimetic considerations on energy technologies it became clear that an abundant supply of cheap hydrogen is the real bottleneck for a future worldwide sustainable energy economy. Today hydrogen is produced from methane. Photovoltaic and wind electricity for water splitting to obtain hydrogen is still too expensive for commercial applications. If cheap hydrogen would be available, affordable sustainable carbon based fuels and chemicals could also be produced. A key advantage of such a fuel technology would be that all our present fossil fuel infrastructure could be maintained. This would be an enormous economic and strategic advantage. The energy infrastructure would, on the other hand, have to be changed if a pure hydrogen economy would be introduced, which would also be possible and interesting. Nevertheless, hydrogen is a technically favoured energy carrier. It is easy to handle as a gas and it is environmentally friendly. In the case of a pure hydrogen economy, however, a parallel carbon based fuel cycle would additionally be needed to supply chemical industry with carbon containing chemicals.

A recent study was published on mechanisms for solar induced hydrogen liberation from water (Tributsch, 2008). There is still significant research needed, but prospects are remarkable for this technology. Artificial model systems based on two photovoltaic cells in series for water splitting have yielded a solar energy conversion efficiency of 18 % for hydrogen generation (Licht et al., 2000). This is 36 times higher than the 0.5% efficiency reached for biomass production via three harvests during one year of sugar cane in a tropical agricultural region. And artificial light-induced generation of hydrogen would not require fertile land. Methods of photo-induced solar hydrogen generation have definitively a future.

But technologies using direct solar light for hydrogen generation are presently much to expensive because solar energy influx has a low density, 1 kW/m^2 at noon, which has to be further reduced by a factor of 5-7 because of the sun movement and the day-night cycle (to an averaged influx of solar energy of only 142-200 W/m^2). Commercial photovoltaic devices today only convert 10-15% of this energy. Wind energy systems may work day and night with favourable technical efficiency and typically operate in energy density ranges between poor wind conditions of 150 W/m^2 to good ones of 350 W/m^2 and excellent wind conditions of 500 W/m^2.

Wave energy technology is based on utilization of kinetic energy of water, which has an 800 times higher density than air. A wave of 3 m can therefore supply an energy density of 36 kW per meter of wave crest. A 6 m high wave already yields an energy density of 180 kW/m. This explains why wave energy is a potentially much more economic sustainable energy source than wind (fig. 1). Smaller and thus cheaper devices can be applied for energy harvesting. Compared to wind energy wave technology is however not yet a mature technology. It is still to be located at the beginning of a longer learning curve. Different function principles for wave energy harvesting devices are still being explored and compared. They are typically based on oscillating water columns, multi-body hinged devices, or overtopping systems, and the installations are typically located at or near the coast to deliver energy directly via electrical cable. Problems, such as matching the resonant frequency of wave energy device to wave frequency, efficiency optimization, survivability of devices and materials in agitated sea water are still being studied. There is a lot of valuable information available in the literature on wave energy and its technology (e.g. Cruz, 2008; McCormick, 2007).

The present contribution does not aim at our present wave technology, but at identifying and discussing the most efficient and potentially cheapest hydrogen production technology via wave energy far away from coast areas in stormy oceans. Ocean regions with the highest possible energy density for wave technology will be considered in order to explore its technical and economic potential. Hydrogen generation from waves in stormy seas far away from ocean shores has, to our knowledge, not yet been considered as a technological option. But such an extreme technical situation of a stormy ocean has been selected here as a probe towards a massive new generation strategy for sustainable and economic fuel.

2. The potential of open sea wave energy

The total solar power incident on the earth has been estimated to amount to approximately 174.000 TW (10^{12} W = 1TW = terawatt). Sometimes a value of 121.000 TW is given, which considers that approximately 30% of the radiation is again reflected into space and not absorbed on the earth surface (fig. 1). For comparison, energy consumption by man presently amounts to 15 TW, with 2 TW accounting for electrical energy. This amount is more than 8000 times smaller than the solar energy absorbed on earth. By the end of the century mankind is however expected to consume approximately 45 TW, with 6 TW accounting for electric energy. Years ago it has been estimated by geophysicists that approximately 2% of the solar power absorbed on the earth surface is converted into mechanical energy of wind, waves and ocean currents. This would amount to approximately 2420 TW. Maybe up to 0.6% of the absorbed solar power could be converted into waves. This rough estimate yields approximately 726 TW, much more than mankind consumes. The Open University in GB teaches in an introduction to energy sources, that a global wind power of as much as 10.000 TW (5.7% of the incident solar energy) could be available. The global wave power was estimated at 1000 TW. (Openlearn, 2011). A power share of 300 TW is assumed to be available as kinetic energy on the earth surface by Twidell and Weir (2006). Lueck and Reid estimated the downward atmosphere-ocean mechanical energy flux to 510 TW. However not more than 10% of the energy, 51 TW, is expected to enter the ocean (Lueck and Reid, 1984).

These relatively high values for the global wave energy potential have to be confronted with specialized studies on availability of wave power. In 1976 Panicker estimated the resource of wave energy in ocean waters with a depth of more than 100 m to approximately 1-10 TW. Also Issaacs and Seymour (Isaacs and Seymour, 1973) limited the global wave power potential to only 1-10 TW. This order of magnitude was recently confirmed by a quite elaborate study based on evaluation of a huge set of data on waves from satellite altimeter and buoy data (the WorldWaves data base, Topex/Poseidon (1992-2002), Jason-1 (2002-2006)) (Mork et al., 2010). These authors estimated a global wave power potential of only 3.7 TW. The World Energy Council appears to confirm that order of magnitude by assuming a wave energy potential of approximately and exceeding 2 TW. A potential of 2 TW has already been estimated just for the global coastal wave power potential (Previsic, 2004). The wave power potential of the open ocean must be much larger, not only because of the much larger area but also because of much higher waves and all year round wave activity close to arctic and antarctic regions (Topex Poseidon radar data).

Compared to the 121.000 TW solar energy input an estimate of 3.7 TW global wave power appears to be rather modest (a fraction of only 1.65 10^{-5}). This can be shown with some very simple considerations. If 3.7 TW are divided by the surface area of world oceans (361.2 million square kilometres) a medium wave energy of only 0.01 W/m^2 results. If 3.7 TW would be distributed on, let us say, 5% of the sea surface, where 4-6 m waves are known to occur practically all year around (Topex/Poseidon data), an average areal energy density of only 0.2 W/m^2 would be obtained. Such waves would barely be visible – in contrast to the large waves in huge areas on the open sea, such as in the roaring forties.

The global wave power of 3.7 TW reported appears to be too low compared to typical near-coast wave power of 30-40 kW/m or up to 200 kW/m in a stormy open sea. A 3 meter high wave has typically a wave length of 10 m and appears with a frequency of 0.2 Hz. Its power per crest length of 30-40 kW/m would have to be multiplied by 0.13 s to reach a much lower areal wave energy density of 3.9-5.2 kWs/m^2 (see formula below). The power, per m^2 of surface area, present in form of kinetic and potential energy of the wave would thus be 3.9-5.2 kW/m^2. Assuming a 4 kW/m^2 energy density of a wave field, 250 km^2 would already add up to 1TW. For a commercial harvest of such an amount of power maybe an area of 1000 km^2 would be needed. As the global wave power distribution on the world map (for example fig. 2 of Mork et al, 2010) visualizes, immensely much larger areas show measured wave power of 30-40 kW/m (crest length) or higher. It is therefore not clear, why a global wave power potential of only 3.7 TW is obtained.

The low estimated wave energy potential in some studies also contradicts a comprehensive estimate of wind power over land and near-shore, which yields a power of 72 TW (Archer and Jacobson, 2005). Wind power over the ocean should be significantly higher.

Other approaches to estimating global wave power consider the wind energy input by calculating energy transfer to the ocean surface with theoretical models (Teng et al., 2009). With a wind energy input of 57 TW, determined for 2005, the energy dissipation in deep sea water was calculated to 33 TW, 58% of the wind energy input. Since the ocean surfaces are much larger than the land surfaces, the estimate of 57 TW may be too low compared to the 72 TW estimated for over land and near shore wind potential, cited by Archer and Jacobson (2005).

From this discussion of available information on global wave energy it may be concluded that insufficient and in part contradictory information exists and more experience is needed to reach an objective picture on the global potential of open sea wave power. On the basis of above given information and arguments it is our understanding that open ocean wave power may be of the order of 10-50 TW, much higher than the 3.7 TW estimated by Mork et al. (2010). Regardless the contradictory estimations of global wave power it can be concluded, that a significant portion of mankind's power consumption may be derived from such a sustainable energy source. It can be considered to be a concentrated secondary solar energy, mediated by inhomogeneous heat generation and evaporation processes in the earth biosphere.

3. From near coast to open sea wave energy harvesting

Wave energy technology today is an on-shore or near shore technology, which converts mechanical energy of waves into electrical energy. This electrical energy is typically supplied to the land via an electrical cable. Wave energy is, as already mentioned, still an emerging technology which is still searching for the most adapted technical devices. It is also facing problems, such as material durability and destruction of technical installations in unusually big storms. A limitation of this technology is also the choice of coastal sites, which may compete with other interests and applications. Another one is a reduced height of waves, due to typically flat water near the coast. In populated coastal regions such as in Europe, or Central and North America, where wave power plants are presently installed and operated, the weather is typically such that larger waves are only present during part of the year.

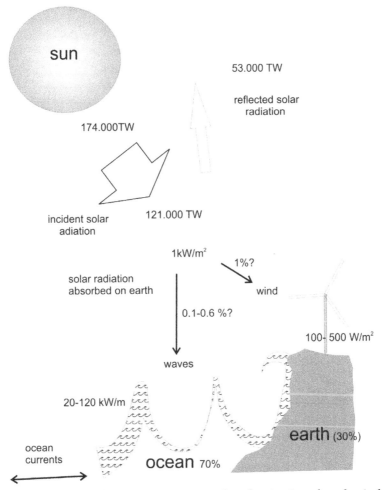

Fig. 1. Scheme visualizing solar energy input on earth and activation of mechanical energy with specification of energy densities encountered for energy harvesting

In order to get the most out of wave energy one should focus on ocean areas with very high waves, which, in addition, are present all over the year. As Topex/Poseidon and Jason-1 radar data show, such areas exist mostly north of the Antarctic continent in the roaring forties, south of Africa and Australia as well as near the southern tip of South America. Such areas of permanent intensive waves also exist close to the arctic circle. At present, there is no ocean wave project known to the author, which is aiming at energetic exploitation of these remote stormy oceans. When changing from a 3 m to a 6 m wave the energy density is increasing five-fold, but simultaneously the wave frequency is decreasing from approximately 0.2 Hz to 0.1 Hz. The resulting energy increase by a factor of approximately 2.5 has additionally to be multiplied by a factor of at least two because of the round the year wave presence in stormy oceans, in contrast to most present sites of wave energy exploitation. The approximately fivefold total increase of energy density is a good precondition for the economic feasibility of this high wave technology approach, as compared to conventional wave energy exploitation, besides of the availability of huge ocean areas for energy harvesting. But the key challenges that remain to be addressed are the survival of the wave technology in the highly stormy environment and the energy transport to industrial destinations.

In order to discuss these subjects it should be helpful to get first acquainted with some essential properties of waves as depicted in fig. 2. A water wave is a periodical phenomenon with very special behaviour. It is typically generated by wind due to the friction between air and the water surface. The mechanism of water wave generation is highly non-linear and very complex (e.g. Johnson, 1997; Falnes, 2002). Waves can develop highly divers pattern. They are, to a large extent, a far from equilibrium phenomenon and part of their properties can only be adequately understood on the basis of dynamic self organization concepts.

The distance between through and crest of a wave, as shown in fig. 2 is the wave height, while the typically used significant wave height is the average height of the highest one third of the waves in a system. The wave length is the distance between two crests, and the period T the time passing between two passages of crests. This period changes with wave height. It approximately triples when the wave height increases from 0.5 to 6 m. Simultaneously the wave length, the distance between two crests, increases from approximately 15 m to 110 m. When waves are passing, this observed kinetic movement of waves, 5 m/s to 13 m/s for the above two situations, is accompanied only by elliptical movements of water molecules (fig. 2). These movements become smaller towards the depth of the sea and practically stop in a distance of approximately a half wave length from the sea surface. In detail, a water particle moves slightly forward with the rising front of the wave, backward with the falling back side of the wave. All together the water molecule at the top of the wave only slightly moves forward as the wave passes. For 5 m high waves, which generate a power of 125 kW/m of crest length (see below), the water movements cease in approximately 45 m depth. The power obtainable from waves per unit of wave crest length is proportional to the square of wave height (Fig. 2) and follows the following formula (P = wave energy power per unit of wave crest length, ρ = water density, g = acceleration by gravity, H = significant wave height, T = wave period)(e.g. Dean and Dalrymple, 1991; Komen et.al, 1994)

$$P = \frac{\rho g^2}{64\pi} H^2 T \tag{1}$$

For a 5 m wave this power per unit crest length yields 125 kW per meter of crest length.

In order to calculate the energy density E contained per unit horizontal area, we have to calculate the sum of kinetic and potential energy of the wave:

$$E = \frac{1}{16} \rho g H^2 \tag{2}$$

Inserting (2) into (1) yields the following relation in the dimension of kWs/m²:

$$E = 1.28 \frac{P}{T} \left[\frac{kWs}{m^2} \right] \tag{3}$$

A period T of 10 s can reasonably be assumed for 5 m waves, changing relation (3) into E = 0.128 P (kWs/m²).

Taking the energy density per unit time, seconds, E/s yields the power of the wave per horizontal unit surface area. When the power, contained in a 5 m wave pattern is calculated per square meter of ocean, a value of approximately 16 kW/m² is thus found. This power can be present day and night, while the maximum solar power (at noon) is only 1 kW/m², and for a day-night average has to be divided by an additional factor of at least 5, yielding only 200 W/m². The average power content in 5 m sea waves is therefore exceeding the average available power in solar intensity by a factor of 80. This shows how much the power of weak sunlight, which is incident on earth, is finally concentrated via weather activity into the power of high waves. Via photovoltaic devices sunlight can only be harvested with an efficiency of now typically 15 %. Wave energy technology, on the other hand, aims at more than 90 % energy conversion efficiency. This may give another factor of 6 when comparing incident solar light energy with the energy present in high waves. Multiplying the 80 fold energy density per square meters of 5 meter waves by a factor of 6 yields a theoretically 480 times higher efficiency for energy harvesting.

Having estimated the power density, per horizontal area, of 16 kW/m² of a 5 m wave field we can now calculate how big the area will be that contains wave energy of one terawatt (1 TW=10^{12} W). This ocean area is 60 square kilometres only. Since only a fraction of that energy can realistically be harvested, a one terawatt (1 TW) seaborne wave power field may cover an area of 200 to 1000 square kilometres (sqkm). Let us now consider that the total surface area of the oceans in the world covers 3.6 10^8 sqkm. The violent oceans with 4-6 m waves, the southern parts (roaring forties) of the Pacific, Indian and Atlantic ocean, a large portion of the Southern Ocean around the Antarctic continent, as well as significant portions of ocean around the arctic ice cover may contribute up to 5 % of the ocean surface. That is, ocean areas with 4-6 m waves (average 5 m), which may prevail at least for a large fraction of the year, may amount up to 18 .10^6 sqkm (this can, for example, also be verified by looking at fig. 2 of Mork et al. (Mork et al, 2010) and estimating the area covered by wave power grid points with more than 40 kW/m of crest length). If 3.7 TW are divided by such an area, an average power per crest length of only 1.56 kW/m is found, much less than the known wave power of 125 kW/m of crest length for 5 m high waves. A wave power of 1.56 kW/m corresponds to a wave height of the order of only 0.8 m. Such a simple estimation again puts a question mark to the elaborate global wave power estimation of only 3.7 TW of Mork et al. (Mork et al., 2010). The global wave power must be much bigger.

These thoughts justify the interest and promise in addressing the challenge of harvesting wave energy from stormy seas. It is the sustainable energy with the highest energy density on earth and it is abundant (excluding sporadic heavy storms, such as hurricanes). And it is localized in remote unpopulated regions of our world. Our energy hungry world would have a realistic opportunity to tap massive sustainable energy from stormy oceans, if it develops a serious effort to build up an adequate technology for harvesting it.

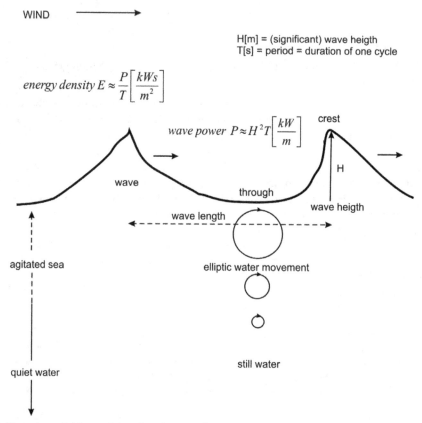

Fig. 2. Drawing which explains the physics of water waves with elliptically circulating water molecules and decreasing movements towards the depth.

4. Challenges and opportunities in stormy seas

As already mentioned, our present, emerging wave energy technology is being developed for coastal or near-coast applications. If wave technology is projected for the stormy, open seas, several boundary conditions will change (Fig. 2). One is that in the open sea, with an average depth of 2000 to 4000 meters the wave energy installations will typically not be anchored on the ground. They will float and may be positioned via satellite and an inbuilt propagation system that uses energy gained from wave power. The second important change will be that the essential wave technological equipment will be placed under water where the wind driven wave activity is not any more felt. This will protect the technology from violent storms. Another important difference will be that the wave energy installations will be so far away from a coast that electrical cables will not be practical and economic. An energy carrying fuel, hydrogen, will directly be produced via electrolysis of sea water on site for transport in gas carrying ships or via undersea pipelines. There will be another difference to conventional near-coast wave energy installations. The fertility of open seas is known to partially be limited by the lack of structures to anchor or harbour marine species. Sunken ships have been known to become artificial reefs with abundant sea life. Large scale open sea wave energy installations may, if properly constructed, become such artificial reefs with abundant opportunities for fish farming. Such a secondary application may have a very favourable impact on economics and may, in addition address a very important emerging problem of mankind. The fish supplies of the oceans are dwindling and world population may approach 10-12 billion people by the end of this century. Developing large scale fish farming on the open sea may be a useful investment.

It will not be practical to construct wave energy installations on site in stormy open seas. Therefore, they should be constructed and finished on assembly lines on far away shipyards in order to be dropped in the sea to be assembled to a large scale energy field. Figure 3 shows and explains the structure and function of such buoys serving as elements of a large energy harvesting field. They transfer the movement of the sea waves via the mechanical forth and back movement of a piston to a large floating underwater body, where electricity is generated for hydrogen generation from water.

In order to deal with stormy oceans it is necessary to gather reliable information on wave patterns and behaviour. Waves in the roaring forties, for example, often reach 9 to 12 m high. Trains of waves of different height and shape arising from different wind patterns may combine in infinite ways. Crest and through may cancel each other and the height of a wave on one place will become equal to the combined heights of waves occurring at that place and time. As a consequence, even 15 to 24 m high waves have occasionally been observed (Hubbard, 1998).

What wave energy exploitation systems could be employed in stormy oceans? Around 65 wave energy devices are listed by Wave Energy Centre (wavec, 2011), a non profit organization for the advancement of wave energy technology. Among them are wave energy harvesting systems such as Pelamis (Ocean Power Delivery Ltd/UK), Powerboy (Ocean Power Technologies/USA), Wave Dragon (Wave Dragon ApS/Denmark), Wave Roller (AW-Energy Oy/Finland) and OEbuoy (Ocean Energy Ltd/Ireland). They include technical principles such as Attenuator Systems, Oscillating Water columns, Overtopping Devices, Oscillating wave devices, axisymmetrical point absorbers, or submerged pressure differential devices. The Pelamis system, for example, which is presently being installed in a commercial wave energy installation in Portugal, is a semi-submerged floating structure composed of cylindrical steel sections which are linked by hinged joints that contain hydraulic pumps. Oil under high pressure drives motors that power electricity generators.

Not considering the fact that hydrogen and not electricity should be the energy product from wave energy fields in stormy seas, no such system will probably survive for a long time, when the wave energy device is floating on the water surface. Powerful waves (e.g. 15 m waves with a period of 15 seconds have a power of 1.7 MW/m of crest length) will earlier or later damage or destroy them. Most conventional wave energy harvesting devices can therefore be excluded from operation in violent oceans. High waves in the open ocean should preferentially be harvested for energy with vertically moving buoys, which are positioning their essential structural and functional elements under the sea surface in the quiet water region.

This quiet region will be approximately half a wavelength below the sea level. Only the swimmer, which is following the periodic wave pattern, will approach and reaches the water surface. Such a strategy places essential existing technology outside the reach of storms. The conversion of mechanical energy via a linear generator into electricity for hydrogen evolution from water should be based on an as simple technology as possible. Delicate motors, pumps and hydraulic systems should be as much as possible avoided. Only very simple and durable technology should be applied, because service interventions on a stormy sea will be a significant challenge.

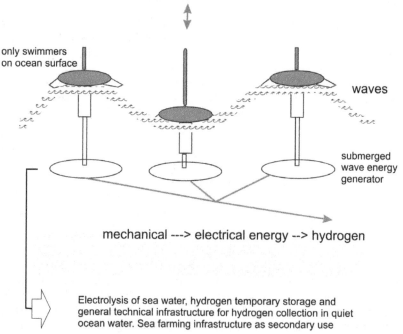

Fig. 3. Structure and function of buoys designed to harvest energy from high waves in stormy seas.

An important construction idea is to drop the buoy in such a way into the sea as to keep it afloat via the swimmer, its floating component. The main submerged body of the buoy should have such a high inertia, and additionally a high water resistance, while being structurally linked to neighbouring devices, to keep it largely immobile and reasonably stable positioned under water. Wave activity above is periodically moving the swimmer up and down while the submerged energy converting and hydrogen generating counter structure will remain largely immobile.

The assembly of many buoys to a buoy field and its function is visualized in fig. 4. The linking together of so many structural and functional wave energy elements will provide the wave power installation with properties comparable to a submerged island over which the trains of water waves are drifting. It may be eventually positioned and propelled with an inbuilt transport and satellite navigation system, supplied by energy generated from waves.

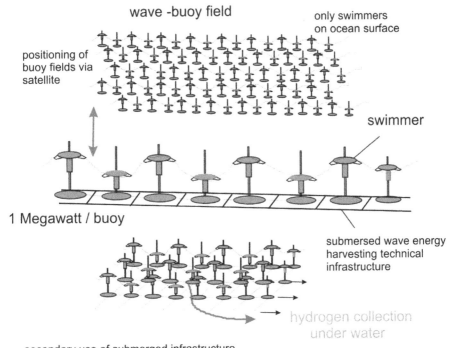

Fig. 4. Wave energy harvesting buoys could be built in mass production and assembled to large fields which could be kept in place or moved via satellite navigation. Besides of the swimmers all technical infrastructure is located in the quiet underwater region. It could also be used in a secondary way for sea food farming installations.

5. Technical infrastructure of open sea wave energy installations

When talking about mechanical under water structures and about difficulties to operate and maintain them, one should first recall what mankind has accomplished in this field. The superpowers operate fleets of sophisticated submarines. There are all kind of diving and sea floor exploring vehicles. Oil and Gas companies are prospecting for fossil fuel, constructing and maintaining production wells in many ocean regions. Under sea channels link continental Europe with Great Britain and they are found in many other geographical locations. Ship technology has no mayor problems with stormy oceans. There is no technical excuse that could make appear a stormy ocean wave energy project a too big technical challenge for our industrial society.

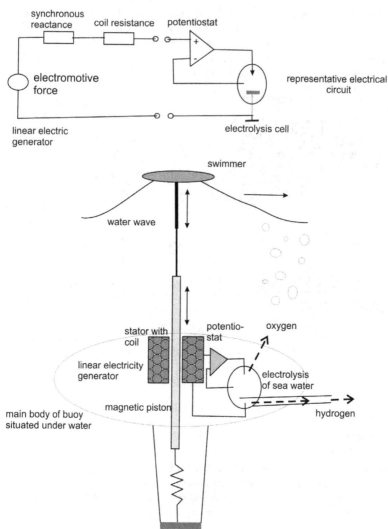

Fig. 5. Technical infrastructure of wave buoy with linear electric generator for direct hydrogen generation from sea water. Above: Representative scheme for electric circuits including linear generator, potentiostat and electrolyser. Below: structural-mechanic set-up of buoy.

Nevertheless, buoys for high waves should be constructed as simple and technically as elegant as possible, because they should be exposed to a violent and corrosive seawater environment and servicing them would appear to be a mayor challenge and would be costly. A vertically, or better, slightly eccentrically moving buoy (because of elliptical water trajectories) should provide the simplest approach. Mechanical wave induced periodical movement could then be directly converted into direct current electricity using a linear electric generator (fig. 5). It directly converts a back and forth motion into electric energy. This is a quite well known principle and is based on the relative movement of an electrical coil and a piston of magnetic material. Is it possible to design a linear, low speed permanent magnet generator with an inherent low load angle and a satisfactory efficiency? The answer is positive and was, for example, given in a theoretical study on applicability of such an electricity generating concept to wave energy harvesting (Danielson, 2003). These calculations show, that it is possible to construct a permanent magnetic generator of 10 kW and 78,6 % efficiency with a magnetic piston moving at a speed of only 0.67 m/s. The load angle, which is relevant for efficiency was found to be satisfactorily low (10.3°). Most losses are due to heat generation through induction in copper, less in iron. Significant improvements appear still to be possible.

Another possibility to drive a linear electrical generator via sheltered underwater technology is the use of a submerged, gas filled vessel which expands and contracts under the varying ambient pressures produced by above passing waves. This principle is known as the "Archimedes Wave Swing" (waveswing, 2011). This contraction and expansion can be transmitted to a linear electricity generator. We consider this technology as more sensitive, since a gas space of changing volume has to be maintained under water for prolonged periods. In addition the efficiency will be limited via the constraints of an expanding air containing vessel, and the infrastructure will be more expensive. The linear generator found already other applications in the linear generator point absorber (Pointabsorber, 2011) and the Ocean Wave Energy Web system (oveco, 2011).

Fig. 6. At the coast wave formation and propagation is frequently affected by geological features of the environment. The transition from a near coast wave energy technology to a deep sea violent ocean wave energy technology will multiply the available power and open up huge ocean areas for sustainable energy harvesting. Instead of electricity hydrogen will become the energy carrier.

All complex electric-electronic infrastructure to produce electricity of well defined quality for hydrogen evolution from sea water in a linear generator wave buoy should be avoided (fig. 5). Instead, the direct current from the linear generator should immediately be provided to a simple potentiostat, that controls the electrical parameters of the electrolysis cell, and passed through suitable cathodes for hydrogen production from sea water (see also below). The hydrogen would be produced under pressure of several bars, depending on the depth in which the water electrolyser is functioning. An under water hydrogen collection and distribution system would finally conduct the hydrogen to a large storage system such as a properly designed submarine vessel. From here the hydrogen could finally be transported away by ship or under-sea pipeline. Today large quantities of methane are transported this way over long distances (200 ships for transport of natural gas) and the economic conditions should not be much different for hydrogen.

The proposed reasonably simple technological route to generate hydrogen from wave energy in deep and violent oceans (fig. 5) justifies mayor efforts towards optimization of technical parts and realization of a pilot installation. As compared to waves near the coast (fig. 6), the technical parameters for energy harvesting will drastically change. By positioning most technical infrastructure in deep still water, where it floats, it will easier survive the impact of storms. And hydrogen instead of electricity transport will secure energy recovery from wide ocean regions.

6. Sea water electrolysis for hydrogen

The direct current electricity produced by the linear generator can be conditioned and transformed to be used in an electrolysis cell for decomposition of sea water. Industrial electrolysis of seawater for hydrogen generation is not an entirely mature technology. When electric current is passing through water, hydrogen is, as well known, liberated at the cathode and oxygen at the anode. The hydrogen evolution process is a simple electrochemical reaction, which does not require noble metals like platinum as an electrode. Nickel is well suited for technical applications. But oxygen evolution is a demanding four-electron transfer process. Special catalysts are necessary, which chemically bind water species during the electrochemical reaction. Typically, transition metal compound are needed to accomplish this. A quite efficient catalytic electrode is RuO_2. In presence of NaCl in the water it also efficiently evolves chlorine, which is technically applied. The problem for seawater electrolysis is that sodium chloride is dissolved in seawater and that the electrode potential for evolution and liberation of chlorine gas at the anode is not very distant from the potential of oxygen evolution from water. If conditions are not properly adjusted poisonous chlorine gas may thus be liberated in a side reaction. While the electrochemical potential for oxygen and hydrogen evolution is changing with the pH value by 0.059V per pH unit, the potential for chlorine evolution does not. It is situated at $E° = +1.36$ V(NHE = normal hydrogen electrode). Sea water has typically a pH value between 7.8 and 8.4 and the oxygen evolution potential is therefore approximately at $E° = +0.758$ (NHE). The hydrogen potential is found at $E° = -0.5472$ V (NHE). Some complication arises in that in non buffered sea water hydrogen evolution is accompanied by a pH shift towards alkalinity (relation (5)), oxygen evolution by a pH shift towards acidity (relation(4).

$$H_2O \rightarrow \tfrac{1}{2} O_2 + 2H^+ + 2e^- \tag{4}$$

$$2H_2O + 2e^- \rightarrow H_2 + 2OH^- \tag{5}$$

The observed pH shifts are, of course, increasing with increasing current density and decreasing with increasing agitation of the seawater electrolyte. Already after the first modern energy crisis thirty years ago J.O.M. Bockris (Bockris, 1989) in his book: "Energy Options" has precisely specified the choice of electrode materials and electrochemical conditions to get efficient hydrogen generation from sea water. If the electrodes are properly selected and adjusted respectively, then chlorine can largely be excluded. This can especially be achieved by choosing highly selective catalysts for oxygen evolution. When, for example, a manganese oxide containing material is used, then oxygen evolution at high current densities can be achieved with 99.6% oxygen evolution and less than 0.4% chlorine evolution (Izuma et al., 1998). When manganese oxide was mixed with molybdenum oxide almost 100% oxygen evolution efficiency was obtained even at high current densities of 1000 A/m^2 (Fujimura et al., 1999). Such selective materials for oxygen evolution should be intensively studied to obtain optimal electrochemical behaviour for the generation of pure hydrogen.

In this connection it is interesting to note that the microscopic and macroscopic algae in the oceans had to overcome exactly the same technical electrochemical problem with respect to chlorine evolution. They also aimed at liberating oxygen from sea water, which involves a complex 4-electron extraction from water, but had to suppress chlorine evolution, which is a kinetically much simpler electrochemical process. Nature obviously succeeded since there is no chlorine evolution problem in the photosynthetic process. And it succeeded using a molecular $CaMn_4O_x$ catalysis complex, which contains manganese, as the above mentioned technical catalyst.

Sea water also contains a series of ions, of which Ca^{2+} and Mg^{2+} can produce problems at he hydrogen evolving cathode. Proton consumption there generates alkalinity (compare relation (5)), which may lead to precipitation of $CaCO_3$ and $Mg(OH)_2$. A high turnover of fresh seawater may avoid this phenomenon during electrolysis on the open ocean.

While some more industrial progress and technical experience will be required in the field of seawater electrolysis, the main parameters appear to be under control. The technique can be handled, provided adequate medium term efforts in research and technology are initiated. A mayor practical challenge will be to build an electrolysis unit, that is sufficiently robust to work unattended and automatically for a very long time. The strategy to make also this unit as simple as possible may pay off.

7. The challenge of under water equipment corrosion and fouling

It is well known that under seawater structures and interfaces are subject to various deteriorating phenomena. They range from electrochemical and bacterial corrosion to inorganic and biological fouling processes. The last two phenomena comprise consequences of processes, during which inorganic or organic deposits form on under seawater structures. Inorganic deposits may form as a consequence of precipitations and may induce solid state and electrochemical reactions of degradation of structural material. Bio-fouling may be induced by barnacles, mussels and snails, which stick to structural under water parts and damage them by chemically degrading them. Barnacles, for example, produce an epoxy-like glue and may even stick to Teflon. Material degradation may occur through the glue of attachment or through products of metabolism or through acids used by the organisms to condition interfaces. Traditional defences against fouling were toxic paints, fouling resistant materials or periodic mechanic cleaning procedures. Copper ions were found to keep barnacle and mussel larvae from settling. They can artificially be generated via copper anodes by applying electricity.

The skin of sharks is covered by small scales or teeth like structures so that the shark surface feels like sand paper. These scales are covered by grooves, which are oriented parallel to the propagation direction. They appear to decrease the water resistance, but also suppress the settling of marine organisms. The explanation is that the highly structured skin of sharks with scales that also flex against each other does not provide reliable anchoring areas for marine organisms. Other sea animals defend themselves against attachment by other organisms by generating slimy interfaces, which are also unsafe anchoring grounds. Sea organisms seem to distrust a slimy underground and tend to avoid it. Sea mammals like whales have a very smooth skin, which provides little roughness for anchoring. In the few remaining tiny cavities there is a gel present, which may additionally discourage the settlement of marine organisms. Among sea animals there is also the strategy to produce a highly toxic interface or frequently scale off and replace the skin. Also specialized small fish have adapted to keep skins of sea animals free of undesired colonists. They have an advantage in sharing prey with their host. There is already a good technical basis for fighting fouling processes on underwater structures and there is also a good chance that biomimetic approaches will finally yield a selection of reliable remedies against this complex problem.

8. The expected impact of cheap hydrogen

Today, hydrogen is nearly entirely produced from fossil fuel (methane) and only when electricity is very cheap, such as at very large or at remote hydroelectric stations, it is generated in a sustainable way. But generation of hydrogen from wind energy and photovoltaics via electrolysis is a process which is today several times more expensive than hydrogen generation from natural gas. This is the biggest obstacle towards a dynamic development of sustainable hydrogen technology. The expensive sustainable electricity generated within and near populated areas is more efficiently used directly. It is for this reason that the here discussed approach towards cheap hydrogen from stormy oceans will be of strategic importance. Only when cheap hydrogen will be available, hydrogen technology will become really attractive (fig. 7). Hydrogen has nearly all ideal properties, which methane has, but burns in addition to water vapour only and not to carbon dioxide, as methane does. It does therefore not negatively affect the environment by contributing to the greenhouse effect. Hydrogen can be transported via pipelines or gas containers like natural gas. When stored, however, the space above has to be ventilated. For this reason, car garages have to be specially built to avoid explosion accidents, the uncontrolled combination of hydrogen with oxygen from the air. Hydrogen is an ideal fuel for transport in down-town areas, because it keeps the air clean, as already amply demonstrated by operating hydrogen buses. But certain preconditions are needed. Because it is so light it is also ideal for air-born transport. It reduces air transportation costs, and, since only water vapour is generated, no damaging atmospheric pollution should be produced. It is well

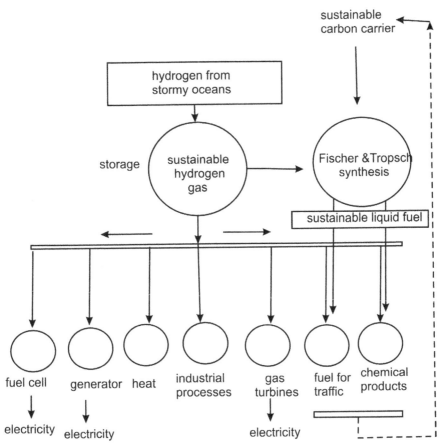

Fig. 7. Scheme explaining the use of cheap hydrogen from stormy oceans. Hydrogen can be directly used as energy carrier, but also added to a sustainable carbon containing carrier (e.g. from gasified biomass) for the production of sustainable liquid fuels

known that fossil fuel combustion by airplanes leads to the emission of tiny carbon containing aerosol nano-particles in the atmosphere. These particles act as nucleation centres for water vapour. Clouds are formed which are dimming solar light incident on our earth. Such a problematic phenomenon could be largely eliminated via hydrogen powered airplanes. Hydrogen could also replace carbon as a reducing agent in metallurgical steel production processes and it could improve and clean up many chemical processes. The advantage of hydrogen compared to carbon is that clean water vapour results instead of polluting carbon dioxide. Hydrogen could also have many applications in daily life. Instead of burning hydrogen in a flame, it could simply be made to combine with oxygen via a catalyst such as finely divided platinum nanoparticles on a porous ceramic structure. It could be used as a plate for cooking or for heating a room. Hydrogen energy could also serve as an important source of clean water in areas where water is scarce. Since hydrogen burns to water vapour, energy turnover is a source of water. For an US American energy consumer 66 litres of water would daily be produced in a side reaction. A European energy consumer would generate approximately 33 litres of water. This is a reasonable amount, considering that the average daily water consumption in Africa is 47 litres, in Asia 85 litres. Water from hydrogen burning could become an important natural resource for population centres with high- energy consumption, as well as for the environment, where water is lacking. One big additional advantage of hydrogen is, that it can easily be converted into electricity via fuels cells. The catalysis of this reaction is comparably simple and efficient.

It has been pointed out that nature is using solar light to split water for hydrogen, but hydrogen is added to a carbon containing carrier, carbon dioxide, so that all kind of energy carriers and chemicals can be produced. On the basis of cheap and abundant hydrogen, our industrial society could follow the same energy strategy as nature. On the basis of Fischer & Tropsch catalysis hydrogen can be added to gasified biomass for synthesis of gasoline and diesel as well as of all kind of chemicals. The big advantage of such strategy is that, on the basis of such sustainable fuels, all our fuel production and distribution infrastructure could remain the same. This would be an enormous financial advantage, considering the vast amount of money, which already has been introduced into transport, conversion and distribution systems for fossil fuel. Simultaneously, however, artificial technologies for carbon dioxide fixation and biomass-generation would have to be developed, because present biomass production is not sufficient for the discussed fuel strategy and competes with food production (Tributsch, 2011).

Such advantages of hydrogen justify all efforts towards its cheap production. Generating it where the secondary solar energy reaches its maximal density (in high wave regions of the ocean), and where a huge sea area is available for modular power plant construction, would seem to be a logic strategy. The elevated energy density of high waves in combination with their availability all over the year guarantees a significant cost advantage, even if higher logistic and maintenance costs would arise.

9. Discussion and summary

Present doubts, after the nuclear disaster in Japan, whether massive nuclear energy technology could safely be handled, motivate a more aggressive development of sustainable energy. This contribution investigated the feasibility of utilizing the most dense and simultaneously most abundant secondary solar energy source, the energy of waves in stormy seas. Compared with the sunlight arriving at low energy density, 5 meter wave areas of an ocean have an eighty times higher energy density which could be harvested with up to 480 times higher energy output. This provides an immense opportunity to produce hydrogen in an economical way. An underwater technology is proposed, which is most basic in terms of infrastructure and simplicity, promising high cost efficiency and durability. A technology is discussed in which only the swimmers of buoy fields reach the water surface to periodically follow the movements of the waves. All sensitive parts as well as the main mechanical structures of the buoy fields should be positioned in deeper quiet water. The stormy wave fields would just drift over them, while periodically moving the swimming buoy elements. Challenges such as seawater electrolysis for hydrogen generation and marine corrosion and fouling of underwater structures were discussed. It is suggested that the proposed technology could be realized and optimized with adequate scientific and technological support within three decades. The cheap hydrogen produced could significantly accelerate the general sustainable development. Sustainably generated cheap hydrogen is a key element for transforming our present mostly fossil energy economy. Added to gasified biomass it could be used to produce sustainable gasoline and diesel with the consequence that all the fossil energy production and distribution infrastructure could be maintained with significant cost advantage. Only the fuel would change from fossil to sustainable. To accomplish such a goal we have to go a courageous step towards energy harvesting from stormy oceans.

10. References

Archer, C.L. ; Jacobson, M.Z.,(2005) *Evaluation of global wind power*, J. Geophys. Res. 110, D12110, doi : 10.1029/2004JD005462.2005

Bockris, J. O`M. (1980), *Energy Options*, Australian & New Zealand Book Company, Sydney, ISBN 0470269154

Cruz, J. (Ed) (2008) Ocean wave energy: current status and future perspectives, Springer, ISBN 0-521-59832-X

Danielson, O., (2003) « Design of linear generator for wave energy plant, Master thesis, Uppsala University, Engineering Physics, http://www.el.angstrom.uu.se/meny/articlar/Design%20of%20a%20linear%20generator%20for%20wave%20energy%20plant4.pdf (retrieved 15.7.2011)

Dean, R.,G., Dalrymple, R., A., *Water wave mechanics for engineers and scientists*. Advanced Series on Ocean Engineering 2. World Scientific, Singapore, ISBN 978-9810204204

Falnes, J., (2002), *Ocean waves ond oscillating systems*, Cambridge University Press, ISBN 0521017491

Fujimura, K. ; Matsui, T. ; Izumiya, K. ; Kumagai, N. ; Akiyama, E. ; Habazaki, H.; Asami,K.; Hashimoto, K., (1999), *Oxygen evolution on manganese-molybdenum oxide anodes in seawater electrolysis*, Mat. Sci. Eng. A 267, 254-259

Hubbard, R. (1998), *Boater`s Bowditch : The small craft Americal Practical Navigator*, International Marine, ISBN 0-07-030866-7

Isaacs, J.D., Seymour, R.D. (1973) *The ocean as a power resource*, Int. Journal of Environmental Studies, vol. 4(3), 201-205, 1973

Izumuya, K. ; Akiyama, E. ; Habazaki, H. ; Kumagai, N. ; Kawashima, A. ; Hashimoto, K., (1998), Electrochimica Acta, 43 (21) 3303-3312

Johnson, R.S., (1997), *A modern introduction to the mathematical theory of water waves*, Combridge Texts in Applied Mathematics, Cambridge University Press, UK. ISBN 0-521-59832-X

Komen, G.J. ; Cavaleri, M.A. ; Donelan, K. ; Hasselmann, S. ; Jansen, P.A.E.M., (1994) Dynamics and modelling of ocean waves, Cambridge University Press, UK. ISBN 978-0521577816

Licht S., Wang B., Mukerji S., Soga T., Umeno M., Tributsch H. (2000), *Efficient Solar Water Splitting, Exemplified by RuO$_2$-Catalyzed AlGaAs/Si Photoelectrolysis,* J. Phys. Chem. B, 104, (2000) 8920-8924

Lueck, R.; Reid, R. (1984) *On the production and dissipation of mechanical energy in the ocean*, J. Geophys. Res. 89, 3439-3445

Mork, G. ; Barstow, S. ; Kabuth, A. ; Pontes, T.M., *Assessing the global wave energy potential*, Proceedings of OMAE2010, 29th International Conference on Ocean, Offshore Mechanics and Arctic Engineering, June 2-6, 2010, Shanghai, China

McCormick, M. (2007), *Ocean wave energy conversion*, Mineola, NY, Dover Publications, ISBN 0486462455

Openlearn, http://openlearn.open.ac.uk/mod/oucontent/view.php?id=399545&section=3.1 , /retrieved 16.7.2011)

Oveco, http://www. oveco.com (retrieved 16.7.2011)

Pointabsorber   : http://www.eureka.findlay.co.uk/archive_features/Arch_Electrical_electronics/f-waves/f-waves.htlm (retrieved 16.7.2011)

Teng, Y. ; Yang, Y. ; Qiao, F. ; Lu, J. ; Yin,X. (2009) , *Energy budget of surface waves in the global ocean*, Acta Oceanologica Sinica, Vol. 28, 5-10

Tributsch H. (2008), *Photovoltaic hydrogen generation*, Int. J. Hydrogen Energy 33, 5911-5930; doi: 10.1016/j.ijhydene.2008.08.017

Tributsch, H. (2011)" *Energy-Bionics: The Bio-analogue Strategy for a Sustainable Energy Future*" in "*Carbon-neutral Fuels and Energy Carriers: Science and Technology*" (N. Muradov, T.N. Veziroglu editors), Taylor & Francis, ISBN 978-1-4398-1857-2

Twidell, T.; Weir, T., (2006) Renewable Energy Resources, Taylor & Francis, ISBN 0-419 - 25320-3

Wave Energy Centre, httw://www.wavec.org/index.php/17/technologies (retrieved 16.7.2011)

10

Design Issues in Radio Frequency Energy Harvesting System

Chomora Mikeka and Hiroyuki Arai
Yokohama National University
Japan

1. Introduction

Emerging self powered systems challenge and dictate the direction of research in energy harvesting (EH). State of the art in energy harvesting is being applied in various fields using different single energy sources or a combination of two or more sources. In certain applications like smart packaging, radio frequency (RF) is the preferred method to power the electronics while for smart building applications, the main type of energy source used is solar, with vibration & thermal being used increasingly. The main differences in these power sources is the power density; for example RF (0.01 ~ 0.1 µW/cm²), Vibration (4 ~ 100 µW/cm²), Photovoltaic (10 µW/cm² ~ 10mW/cm²) and Thermal (20 µW/cm² ~ 10mW/cm²). Obviously RF energy though principally abundant, is the most limited source on account of the incident power density metric, except when near the base stations. Therefore, in general, RF harvesting circuits must be designed to operate at the most optimal efficiencies.

This Chapter focuses on RF energy harvesting (EH) and discusses the techniques to optimize the conversion efficiency of the RF energy harvesting circuit under stringent conditions like arbitrary polarization, ultra low power (micro or nanopower) incidences and varying incident power densities. Harvested power management and application scenarios are also presented in this Chapter. Most of the design examples described are taken from the authors' recent publications.

The Chapter is organised as follows. Section 2.1 is the introduction on RF energy sources. Section 2.2 presents the antenna design for RF EH in the cellular band as well as DTV band. The key issue in RF energy harvesting is the RF-to-DC conversion efficiency and is discussed in Section 2.3, whereas Section 2.4 and 2.5 present the design of DTV and cellular energy harvesting rectifiers, respectively. The management of micropower levels of harvested energy is explained in Section 2.6. Performance analysis of the complete RF EH system is presented in Section 3.0. Finally, conclusions are drawn in Section 4.0.

1.1 RF energy sources

These include FM radio, Analogue TV (ATV), Digital TV (DTV), Cellular and Wi-Fi. We will present a survey of the measured E-field intensity (V/m) for some of these RF sources as shown in Table 1, [1]-[2]. Additionally, measured RF spectrums for DTV and Cellular signals are presented as shown in Fig. 1 to show on the potential for energy-harvesting in these frequency bands. In general, many published papers on RF-to-DC conversion, have presented circuits capable of converting input or incident power as low as -20dBm. This means that, if an RF survey or scan finds signals in space, with power spectrum levels around -20dBm, then, it is potentially viable to harvest such signal power. In Fig. 1 (left side), the spectrum level is well above -20dBm and hence, a higher potential for energy harvesting. In Fig. 1 (right side), while the spectrum level is below -20dBm, what we observe is that the level increases with decrease in the distance toward the base station (BTS). Using free space propagation equation with this data, it was calculated that at a distance 1.4 m from the BTS, the spectrum level could measure 0dBm. An example calculation and plot for the estimated received power level, assuming 0dBi transmitter (BTS) and receiver antenna gains and free space propagation loss (FSPL) for FM and DTV is presented in Section 2.1.1. For the example estimation in Section 2.1.1, we select FM and DTV because they measured with a higher level than cellular and Wi-Fi for example.

Source	V/m	dBm	Reference	
FM radio	0.15~3		Asami et al.	
Analogue TV	0.3~2			
Digital TV	0.2~2.4	-40~0.0	Asami et al.	Arai et al.
Cellular		-65~0.0	Mikeka and Arai	
Wi-Fi		≅ -30		

Table 1. RF energy sources, measured data.

In Table 1, FM radio has the highest E-field intensity implying the highest potential for energy harvesting. However, due to the requirements for a large antenna size and the challenges for simulations and measurements at the FM frequency i.e. 100 MHz or less (See Section 2.2.3, example FM antenna at 80 MHz), this Chapter will focus on DTV (470~770 MHz band) and Cellular (2100 MHz band) energy harvesting.

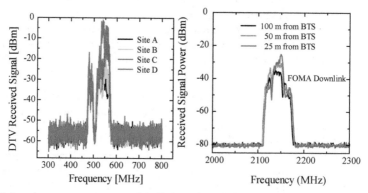

Fig. 1. DTV signal spectrum measured in Tokyo City (left side graph) and Cellular signal spectrum measured in Yokohama City (right side graph).

The received DTV signal power is high and also wide band, presenting high potential for increased energy harvesting unlike in cellular signals. We demonstrated in [2] that the total RF-to-DC converted power is roughly the integral over the DTV band (1), and is significantly larger than in the case of narrow band cellular energy harvesting.

$$P_{DC(DTV)} = \alpha \int_{470}^{770} \delta P_{DC}(f) df, \tag{1}$$

where α is the attenuation factor on the rectifying antenna's RF-to-DC conversion efficiency due to multiple incident signal excitation. δP_{DC} is the small converted DC power from each of the single DTV signals in the 470 MHz to 770 MHz band.

In detail, we derive (1) from fundamentals as follows.

The incident power density on the rectifying antenna (rectenna), $S(\theta,\phi,f,t)$, is a function of incident angles, and can vary over the DTV spectrum and in time. The effective area of the antenna, $A_{eff}(\theta,\phi,f)$, will be different at different frequencies, for different incident polarizations and incidence angles. The average RF power over a range of frequencies at any instant in time is given by:

$$P_{RF}(t) = \frac{1}{f_{high} - f_{low}} \int_{f_{low}}^{f_{high}} \int_{0}^{4\pi} S(\theta,\phi,f,t) A_{eff(\theta,\phi,f)} d\Omega df \tag{2}$$

The DC power for a single frequency (f_i) input RF power, is given by

$$P_{DC}(f_i) = P_{RF}(f_i,t).\eta \left(P_{RF}(f_i,t), \rho, Z_{DC} \right), \tag{3}$$

where η is the conversion efficiency, and depends on the impedance match $\rho(P_{RF},f)$ between the antenna and the rectifier circuit, as well as the DC load impedance. The reflection coefficient in turn is a nonlinear function of power and frequency.

The estimated conversion efficiency is calculated by P_{RF}/P_{DC}. This process should be done at each frequency in the range of interest. However, DC powers obtained in that way cannot be simply added in order to find multi-frequency efficiency, since the process is nonlinear. Thus, if simultaneous multi-frequency or broadband operation like in DTV band is required, the above characterization needs to be performed with the actual incident power levels and spectral power density. In this Chapter, we shall demonstrate DTV spectrum power harvest, given a rectenna than has been characterised in house at each single frequency in the DTV band.

1.1.1 An example calculation and plot for the estimated received power level

In this example we consider Tokyo's DTV and FM base stations (BS) as the RF sources. Both DTV and FM BS transmitter power (P_t) equals 10 kW (70dBm). The antenna gains are assumed 0dBi in both cases but also at the points of reception for easiness of calculation but with implications as follows. Assuming 0dBi antenna at each reception point, demands that we specify the frequency of the transmitted signal. For this reason we specify DTV signal frequency to be equal to 550 MHz while the FM signal frequency equals 80 MHz (Tokyo FM).

The received power, P_r is calculated using the simplest form of Friis transmission equation given by

$$P_r = P_t + G_t + G_r + FSPL ,\qquad(4)$$

where P_t = 70dBm, $G_t = G_r$ = 0dBi. G_r is the receiving antenna gain while FSPL is the free-space path loss given by

$$FSPL(dB) = 20\log(d) + 20\log(f) + 32.45 ,\qquad(5)$$

where d is in (km) and f is in (MHz). The plot for the received power as a function of distance from the DTV and FM base stations is shown in Fig. 2.

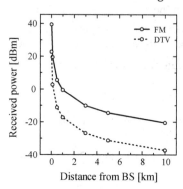

Fig. 2. DTV and FM received signal power level against distance.

With respect to Fig. 2, FM registers higher received power level than DTV at every reception point due to its lower transmit frequency and hence lower free-space path loss. For example at 1 km distance, FM received power level is -0.51dBm while for DTV, the received power is -17.26dBm. The important thing however, is that the received power level is frequency independent. It means that P_t is the transmitter power and the received power level at the position of distance d is $\dfrac{P_t}{4\pi d^2}$. However, if we assume 0dBi antenna at each reception point as in the above example, the power level is different because the antenna size of 0dBi is frequency dependent. As a result, high transmit power level is favorable for RF energy harvesting. Also near the base station is favorable.

1.2 Antenna design for the proposed RF energy harvesting (EH) system

It is well known that RF EH system requires the use of antenna as an efficient RF signal power receiving circuit, connected to an efficient rectifier for RF-to-DC power conversion. Depending on whether we want to harvest from cellular or DTV signals, the antenna design requirements are different. We will discuss the specific designs in the following sub sections.

1.2.1 Cellular energy harvesting antenna design

We propose a circular microstrip patch antenna (CMPA) for easy integration with the proposed rectifier (Section 2.5.1). However, the use of circular microstrip patch antennas (CMPA) is often challenged by the need for impedance matching, circular polarization (CP) and higher order harmonic suppression.

To address the above concerns, we create notches on the circular microstrip patch antenna. In our approach, we use only two, thin, fully parameterized triangular notches to achieve higher order harmonic suppression, impedance matching and circular polarization, all at once. This is the novelty in our proposed antenna. Our proposed CMPA is shown in Fig. 3. We study the behaviour of CMPA surface current vectors when notches (triangles ABC) are created on the structure at $\alpha = 45^0$ and $\alpha = 225^0$.

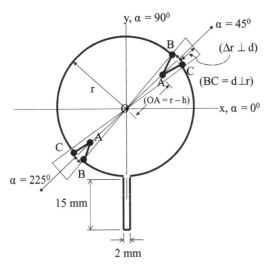

Fig. 3. Cellular energy harvesting antenna structure.

Notch parameters d and h in Fig. 3 were investigated by calculation using CST microwave Studio.

Without notches, the CMPA's input is not matched at f_c= 2.15 GHz as shown in Fig. 4 (left side). However, with notches, matching is achieved. The parameter combination $d = 7$ and $h = 6$ offers a matched and widest band input response and hence we adopt it for cellular energy harvesting applications.

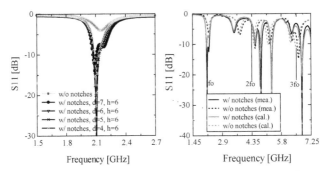

Fig. 4. *Left side:* d and h parameter investigation. *Right side:* Comparison between (cal.) and (mea.) S_{11} at $f_0 = 2.175$ GHz, $2f_0 = 4.35$ GHz and $3f_0 = 6.53$ GHz. The adopted notch parameters are d=7 mm while h= 6 mm.

The comparison between calculated and measured S_{11} is shown in Fig. 4 (right side). The 2nd and 3rd harmonics are suppressed as required by design. The comparison between calculated and measured radiation patterns is shown in Fig. 5, where $E_\vartheta \cong E_\varphi$ due to the 45⁰ tilted surface current vector. Ordinarily, without notches, the surface current vector is parallel to the microstrip feeder axis. In conclusion, our proposed CMPA is sufficiently able to suppress higher order harmonics while simultaneously radiating a circularly polarized (CP) wave. The CP is required to efficiently receive the arbitrary polarization of the incident cellular signals at the rectenna.

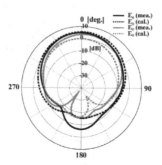

Fig. 5. Cellular energy harvesting antenna pattern at f_c= 2.15 GHz.

1.2.2 DTV energy harvesting antenna design

Unlike in the cellular energy harvesting antenna, the DTV energy harvesting antenna must be wideband (covering 470 MHz to 770 MHz), horizontally polarized and omni-directional.

The proposed antenna is typically a square patch (57 mm x 76 mm) with a partial ground plane (9 mm x 100 mm). The patch is indirectly fed by a strip line (9 mm x 3 mm). The proposed antenna geometry is shown in Fig. 6. The partial ground plane is used to achieve omni-directivity and a certain level of wide bandwidth. To tune the impedance of this antenna as well as to adjust the bandwidth within the target band, a "throttle" with stepped or graded structures is used between the microstrip feed line and the square patch, as shown in Fig. 6 (left side).

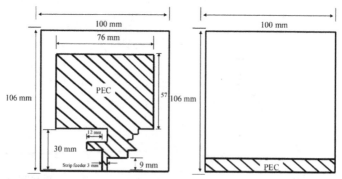

Fig. 6. Proposed DTV antenna geometry. *Left side:* Front view. *Right side:* Back view. The antenna is printed on FR4 substrate; t = 1.6 mm, ε_r = 4.4.

The input response for the proposed antenna is shown in Fig. 7 (left side). The omni directivity is confirmed by measurement at 500 MHz, 503 MHz, and 570 MHz as shown in Fig. 7 (right side). The radiation patterns shown in Fig. 7 are for the xz plane, which happens to be the vertical polarization for the antenna. DTV signals are horizontally polarized and therefore, when using this antenna, the orientation must be in such a way as to efficiently receive the DTV signal. Simply a 90 degree rotation of the antenna along the z axis achieves this requirement.

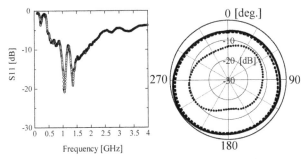

Fig. 7. Proposed DTV antenna performance. *Left side:* The antenna's measured input performance. *Right side:* The omni directivity in the vertical plane is confirmed at 500 MHz, 503 MHz, and 570 MHz. The outermost, black solid and dotted line patterns represent 503 MHz and 500 MHz directivity, respectively. The innermost dotted line pattern is the directivity at 570 MHz.

1.2.3 Example design for an 80 MHz FM half-wave dipole antenna

A half-wave dipole is the simplest practical antenna designed for picking up electromagnetic radiation signals, see Fig. 8 (courtesy of Highfields Amateur Radio Club). Calculating the optimal antenna length to pick up a certain frequency signal is fairly straightforward because antenna physics demand that the total length of wire used in the antenna be equal to one wavelength of the type of electromagnetic radiation it will be picking up. This means that the total length of the antenna should be equal to half the desired wavelength. By converting the 80 MHz frequency into a wavelength, you can thus obtain your antenna length as 1.875m by using the magic equation, $\lambda = \dfrac{c}{f}$. However, the actual length is typically about 95% of a half wavelength in free space, hence a half-wave dipole for this frequency should be 1.788m long, which would make each leg of the dipole 0.894m in length.

1.3 RF-to-DC conversion efficiency improvement techniques

A Schottky diode circuit connected to an antenna is used for RF-to-DC power conversion. To convert more of the antenna surface incident RF power to DC power, high RF-to-DC conversion efficiency is required of the rectifying circuit. Many authors have shown that the efficiency depends on several factors like Schottky diode type, harmonics suppression capability, load resistance selection, and the capability to handle arbitrary polarized incident waves. What is missing in most of these published works is the efficiency optimization for

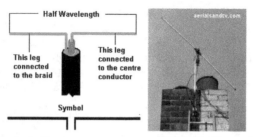

Fig. 8. Half-wave dipole. *Left side:* Antenna structure. *Right side:* Typical deployment.

ultra low power incident waves and the explanation of the physical phenomena behind most of the recommended efficiency optimization approaches.

This Chapter will show for example that a Schottky diode that delivers the highest efficiency at 0dBm incidence may not necessarily deliver the highest efficiency at lower power incidence e.g. –20dBm. We will therefore classify which diodes perform better at given power incidences; of course, this will also be compared to the diode manufacturers' application notes. Simulations in Agilent's ADS using SPICE and equivalent circuit models will compare the performance of few selected Schottky diodes namely; HSMS-2820, HSMS-2850, HSMS-2860, HSC-276A, and SMS7630. Moreover, the effect of the Schottky diode's junction capacitance (C_j) and junction bias resistance (R_j) on the conversion efficiency will be shown from which, special techniques for Schottky diode harmonic suppression and rectifying circuit loading for maximum efficiency point tracking will be presented.

1.3.1 The schottky diode

The classical *pn* junction diode commonly used at low frequencies has a relatively large junction capacitance that makes it unsuitable for high frequency application [3]. The Schottky barrier diode, however, relies on a semiconductor-metal junction that results in a much lower junction capacitance. This makes Schottky diodes suitable for higher frequency conversion applications like rectification (RF-to-DC conversion) [3]. We will demonstrate the effects of junction capacitance and resistance in the following sub section.

1.3.2 The effect of Schottky diode's C_j and R_j on the conversion efficiency

We have studied Schottky diode's C_j and R_j and published our results in [4]. In this work, we designed a rectifying antenna tuned for use at 2 GHz. The circuit proposed in [4] is a voltage doubler by configuration, but we replaced the amplitude detection diode (series diode) with its equivalent circuit adapted from [5]. The results of this investigation show that variation of C_j shifts the tuned frequency position and also introduces a mismatch in the resonant frequency, see Fig. 9 (left side graph). Therefore for this circuit at 2 GHz, we recommend using a Schottky diode having $C_j = 0.2pF$. In general, a smaller value of C_j is desirable at higher frequencies. Similarly, for R_j investigation, a smaller value is desirable for better matching at 2 GHz for example. If the R_j is increased towards 10kΩ, there is a mismatch in the resonant frequency but no shift in the frequency, see Fig. 9 (right side graph). Another approach to the study of Schottky diodes for higher frequency and efficiency rectenna design is presented in [6].

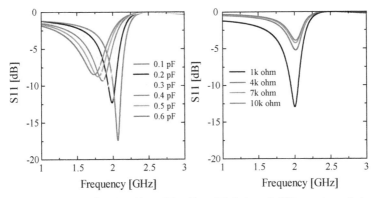

Fig. 9. Schottky diode's C_j effect (Left) and R_j effect (Right) on 2 GHz rectenna's input response.

1.4 Rectifying circuit for DTV energy harvesting

In the design of a DTV energy harvesting circuit, several basic design considerations must be paid attention to. First is the antenna; it must be wideband (covering 470 MHz to 770 MHz), horizontally polarized and omni-directional. Secondly is the rectifier; it must also be wideband, and optimized for RF-to-DC conversion for incident signal power at least - 40dBm. Until recently, very few authors have published on DTV energy harvesting circuit. For the few publications, the antenna could not meet all those three requirements and a discussion on the performance of the harvesting circuit for ultra low power incidences has been neglected. In this Chapter we will present such a rectenna with conversion efficiencies above 0.4% at -40dBm, above 18.2% at -20dBm and over 50% at -5dBm signal power incidence. We will closely compare simulated and measured performance of the rectenna and discuss any observed disparities.

Agilent's ADS will be used to simulate the nonlinear behaviour of the rectifying circuit based on harmonic balance tuning methods. To simulate the multiple incident waves, a multi-tone excitation in the DTV band will be invoked. The wideband input characteristic will be achieved by the input matching inductors and capacitors.

The generic version of our proposed DTV energy harvesting circuit is shown below in Fig. 10. The implementation, however, is in two phases or scenarios as follows. First, we investigate the class called "ultra low power" DTV band rectenna. Secondly, we introduce the "medium power" DTV band rectenna.

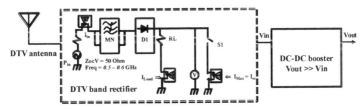

Fig. 10. Generic version of our proposed DTV energy harvesting circuit.

1.4.1 Ultra low power DTV rectenna

We define an ultra low power rectenna as one impinged by RF power incidence in the range between – 40dBm and -15dBm. Below in Fig. 11 is the circuit we designed; optimized for -20dBm input. The matching network is complex so as to achieve a wide band input characteristic. The fabricated circuit was well matched for the frequency range between 470 MHz and 600 MHz. More details about the circuit design can be found in [7].

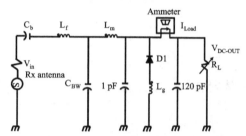

Fig. 11. Ultra low power DTV band rectenna circuit. SMS7630 Schottky diode by SKYWORKS offered the best performance.

The RF-to-DC conversion efficiency for this circuit is shown in Fig. 12 where at input power equal to -40dBm, efficiency is at least 0.4% and rectified voltage equals 1mV; at -20dBm, we have at least 18.2% by measurement and a rectified voltage of 61.7mV. The level of rectified voltage is too low and disqualifies this circuit for purposes of charging capacitors or batteries to accumulate such micropower over time. Instead, boosting the low voltage to usable levels is the option available and we shall discuss this at a later stage, (in Section 2.6).

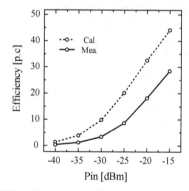

Fig. 12. Ultra low power DTV band rectenna efficiency.

1.4.2 Medium power DTV rectenna

We define a medium power rectenna as one impinged by RF power incidence in the range between – 5dBm and 0dBm. Below in Fig. 13 is the circuit we designed, optimized for -5dBm input. The matching network is simpler than as shown in section 2.4.1 since we require a narrow band around 550 MHz, with received peak power spectrum levels at least -5dBm. The circuit in Fig. 13 is a modification of Greinacher's doubler rectifier. In the circuit, C_b equals 1 pF and is used to block DC current against flowing towards the source. The shunt capacitance, C_{BW} equals 3300 pF and is used to set the input bandwidth. The grounding inductance, L_g equals 56nH (optimal) and is used to improve the RF-to-DC conversion efficiency by cancelling the Schottky diodes (D_b and D_D) capacitive influence; thereby minimizing the harmonic levels (harmonic suppression). We used HSMS2850 diodes in these circuits for their better performance at this level of incident power.

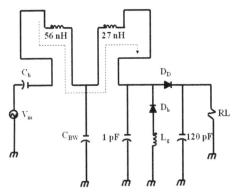

Fig. 13. Medium power DTV rectenna circuit. HSMS 2850 or 2820 from Hewlett-Packard offered the best performance.

The RF-to-DC conversion efficiency for this circuit is shown in Fig. 14 where at input power equal to -5dBm, we achieve at least 50% conversion efficiency by measurement, equivalent to 1.2 V DC rectified at 8.2kΩ optimal load. If we change the load to 47kΩ, over 2 V DC is rectified. This rectenna circuit is ideal for powering small sensors that run on 1.5 V or 2.2 V and draw around 6μA nominal current. If we need to power sensors demanding more power, say at least 2.2 V and 0.3mA to 1.47mA current consumption, we have to accumulate the power in a capacitor over time.

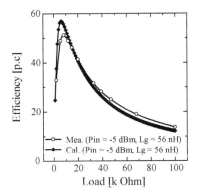

Fig. 14. Medium power DTV band rectenna efficiency.

1.4.3 DTV energy harvesting scenario and application demo

Using the medium power DTV band rectenna, connected to a gold capacitor as an accumulator, energy harvesting was initiated as shown in Fig. 15. Details about the gold capacitor, which include its charge function, backup time and leakage losses are presented in [8]. For the scenario shown in Fig. 15, the accumulated voltage by measurement i.e. capacitor charge function follows the path;

$$V_{acc} = 0.5388\ln(t) + 1.4681 \tag{6}$$

where V_{acc} is the accumulated voltage in volts and t the time in hours. It takes 4.5 hours to accumulate 2.25 V, given a rectified charging voltage and current of 2.4 V and 51μA, respectively, supplied by the DTV band rectenna instantaneously.

With this rectenna, it was possible to power up many different kinds of sensors. Sensors with ultra low power consumption were powered directly, without need to accumulate the power in a capacitor, as shown in Fig. 16.

Fig. 15. DTV energy harvesting in a park at some line of sight from the base station.

Fig. 16. Directly powering a thermometer mounted on a car park wall (right picture). The maximum instant voltage rectification on record equals 3.7 V (left picture).

1.5 Rectifying circuit for cellular energy harvesting

Unlike in the DTV energy harvesting circuit, for cellular energy harvesting, the antenna must be narrowband (50 MHz bandwidth is acceptable), and circularly polarized even though cellular signals are vertically polarized. The circular polarization is desired to maximize the RF-to-DC conversion efficiency of the arbitrary polarization incident signals in the multipath environment. Similarly, the rectifier must be narrowband, and optimized for RF-to-DC conversion over a wide range of incident signal power.

Thinking about the potential applications for cellular energy harvesting is useful. Other authors have reported on powering a scientific calculator or a temperature sensor from GSM energy harvesting. In this Chapter we will present a special application for energy harvesting in the vicinity of the W-CDMA cellular base station and analyze the system performance by calculation from experimental data. A cellular energy harvesting circuit optimized for over 50% RF-to-DC conversion efficiency given approximately 0dBm incidence will be presented.

1.5.1 Cellular band rectenna

Below in Fig. 17 is the circuit we designed, optimized for 0dBm input. Simple input matching network is ideal since we require a narrow band response around 2.1 GHz. The optimum value for L_g equals 5.6nH, where L_g is used to improve the RF-to-DC conversion efficiency as earlier discussed. HSMS2850 diode was used.

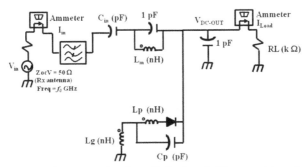

Fig. 17. Shunt rectifier configuration for the cellular band. The matching elements L_m = 3.2nH, while C_{in} =2.5pF. The load resistance is fixed at R_L = 2.1kΩ.

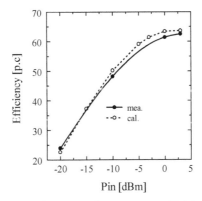

Fig. 18. Conversion efficiency as a function of input power (P_{in}) in the cellular band.

The RF-to-DC conversion efficiency for this circuit is shown in Fig. 18 where at input power equal to 0dBm, we achieve at least 60% conversion efficiency by measurement, given a 2.1kΩ optimal load. This rectenna circuit is ideal for powering small sensors that run on 1.5 V or 2.2 V and 6μA nominal current consumption. If we need to power sensors demanding more power, say at least 2.2 V and 0.3mA to 1.47mA, we have to accumulate the power in a capacitor over time as discussed in section 2.4.3 above.

1.5.2 Cellular energy harvesting application example

Environmental power generation in the neighbourhood of a cellular base station to power a temperature sensor is proposed as shown in Fig. 19 below. Electric field strength measurements in the base station neighbourhood have demonstrated the potential for environmental power generation, and the proposed temperature sensor system is designed based on these values. The rectenna described in Section 2.5.1 is used as the RF-to-DC rectifying circuit with the notched circular microstrip patch antenna (CMPA) proposed in Section 2.2.1. RF-to-DC conversion efficiency equal to 53.8% is obtained by measurement. The temperature sensor made for trial purposes clarifies the capability for temperature data wireless transmission for 20 seconds per every four hours in the base station neighbourhood.

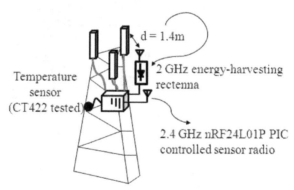

Fig. 19. Application example in the vicinity of the cellular base station.

1.6 Micropower energy harvesting management

A rectifying antenna circuit for -40dBm incident power harvesting generates 1mV at 2kΩ load, given 0.4% efficiency as presented in Section 2.4.1. At -20dBm incidence and at least 18.2% efficiency, 61.7mV is generated given a 2kΩ load [7]. The generated DC power in both of these two cases is in the μW range, hence the micropower definition. To manage such micropower, power accumulation or energy storage is required. Storage devices may either be a gold capacitor, super capacitor, thin film battery or the next generation flexible paper batteries. These storage devices have specific or standard maximum voltage and trickle charging current minimum requirements. Typically, gold capacitors have voltage ratings like 2.7 V, 5.5 V for 100 μA, 10mA or 100mA maximum discharge current. On the other hand, standard ratings for batteries are 1.8 V, 2 V, 3.3 V and 4.1 V. Therefore, to directly charge any of these storage devices from 1mV, or 61.7mV DC is impractical.

Published works have demonstrated the need for a DC-to-DC boost converter placed between the rectifying antenna circuit (rectenna) and the storage device. Recent efforts have demonstrated that a 40mV rectenna output DC voltage could be boosted to 4.1 V to trickle charge some battery. A Coilcraft transformer with turns ratio ($N_s : N_p$) equal to 100 was used in the boost converter circuit. An IC chip leading manufacturer (Linear Technology Corp., LT Journal, 2010) has released a linear DC-to-DC boost regulator IC chip capable of boosting an input DC voltage as low as 20 mV and supplying a number of possible outputs, specifically suited for energy harvesting applications. While this IC is a great milestone, readers and researchers need to understand the techniques to achieve such ICs and also the limitations that apply. In the following sub section, we will describe the methods toward designing a DC-DC boost converter, suitable for micropower RF energy harvesting.

In the design, we will attempt to clarify the parameters that affect the DC-DC conversion efficiency. For this design, Envelope simulation in Agilents's ADS is used. This simulation technique is the most efficient for the integrated rectenna and DC-DC boost converter circuits.

1.6.1 DC-DC boost converter design theory and operation

The DC-DC boost converter design theory and actual implementation are presented in this section. The inequality $V_{in} \ll V_{out}$ defines the boost operation. In this Chapter, our boost converter concept is illustrated in Fig. 20. A small voltage, V_{in} is presented at the input of the boost converter inductive pump which as a result, generates some output voltage, V_{out}. The output voltage is feedback to provide power for the oscillator. The oscillator generates a square wave, F_{OSC} that is used for gate signalling at the N-MOSFET switch.

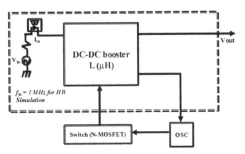

Fig. 20. Boost converter concept.

The drain signal of the N-MOSFET is used as the switch node voltage, V_{sn} at the anode of the diode inside the boost converter circuit block. From the concept presented in Fig. 20, the actual implemented circuit is shown in Fig. 21. The circuit was designed in Agilent's ADS and fabricated for investigation by measurement.

The circuit in Fig. 21 is proposed for investigation. Since a DC-DC boost converter is supposed to connect to the rectenna's output, it therefore, becomes the load to the rectenna circuit. This condition demands that the input impedance of the boost converter circuit emulates the known optimum load of the rectenna circuit. This has the benefit of ensuring maximum power transfer and hence higher overall conversion efficiency from the rectenna input (RF power) to the boost converter output (DC power). In this investigation, as shown in [7], the optimum load for the rectenna is around 2kΩ. In general, emulation resistance R_{em} is given by

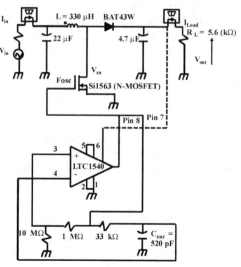

Fig. 21. The proposed boost converter circuit diagram. Designed in Agilent's ADS and fabricated for investigation by measurement.

$$R_{em} = \frac{2LT}{t_1^2 k}\left(\frac{M-1}{M}\right) \tag{7}$$

where L is the inductance equal to 330μH as shown in Fig. 20, $M = \dfrac{V_{out}}{V_{in}}$, T is the period of F_{OSC}, t_1 is the switch"ON" time for the N-MOSFET, and k is a constant that according to [3] is a low frequency pulse duty cycle if the boost converter is run in a pulsed mode and typically, k may assume values like 0.06 or 0.0483. With reference to (7), we select L as the key parameter for higher conversion efficiency while V_{in} = 0.4 V DC is selected as the lowest start up voltage to achieve oscillations and boost operation. Computing the DC-DC boost conversion efficiency against different values of L, we have results as shown in Fig. 22.

From the results above, L = 100μH is the optimum boost inductance that ensures at least 16.5% DC-DC conversion efficiency, given R_L = 5.6kΩ.

Now having selected the optimum boost inductance given some load resistance, the emulation resistance shown in Fig. 23 is evaluated from the ratio of voltage versus current at the boost converter circuit's input.

The results show a constant resistance value against varying inductance. In general, we can say that this boost converter circuit has a constant low input impedance around 82.5Ω. This impedance is too small to match with the optimum rectenna load at 2kΩ. This directly affects the overall RF-to-DC conversion efficiency.

The results show a constant resistance value against varying inductance. In general, we can say that this boost converter circuit has a constant low input impedance around 82.5Ω. This impedance is too small to match with the optimum rectenna load at 2kΩ. This directly affects the overall RF-to-DC conversion efficiency.

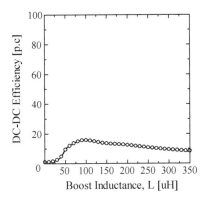

Fig. 22. Boost inductance variation with DC-DC conversion efficiency for a 5.6 kΩ load.

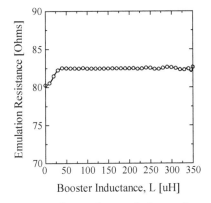

Fig. 23. Boost converter's input impedance: the emulation resistance.

Another factor, which affects the overall conversion efficiency is the power lost in the oscillator circuit. Unlike the circuit proposed in [9], which uses two oscillators; a low frequency (LF) and high frequency (HF) oscillator; in Fig. 21, we have attempted to use a single oscillator based on the LTC1540 comparator, externally biased as an astable multivibrator.

The power loss in this oscillator is the difference in the DC power measured at Pin 7 (supply) to the power measured at pin 8 (output). We term this loss, L_{osc}; converted to heat or sinks through the 10MΩ load. A comparison of the oscillator power loss to the power available at the boost converter output is shown in Fig. 24.

Looking at Fig. 24; we notice that the power loss depends on whether the oscillator output is high or low. The low loss corresponds to the quiescent period where the power lost is almost negligible. However, during the active state, the lost power (power consumed by the oscillator) nearly approaches the DC power available at the boost converter output. This results in low operational efficiency.

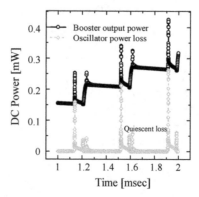

Fig. 24. The power loss in the oscillator.

To confirm whether or not the circuit of Fig. 21 works well, we did some measurements and compared them with the calculated results. Unlike in calculation (simulation), during measurement, L = 330μH was used due to availability. All the other component values remain the same both in calculation and measurement. In Fig. 25 (left side graph) and (right side graph), we see in general that the input voltage is boosted and also that the patterns of F_{osc} and V_{sn} are comparable both by simulation and measurement. To control the duty cycle of the oscillator output (F_{osc}), and the level of ripples in the boost converter output voltage (V_{out}), we change the value of the timing capacitance, C_{tmr} in the circuit of Fig. 21. Simulations in Fig. 25 (left side graph) show that C_{tmr} = 520pF realizes a better performance i.e. nearly constant V_{out} level (very low ripple).

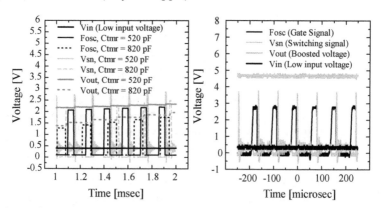

Fig. 25. Voltage characteristics of the developed boost converter circuit. The left side graph represents simulation while the right side graph is for measurements.

Generally, we observe that with this kind of boost converter circuit topology, it is difficult to start up for voltages as low as 61.7mV DC generated by the rectenna at -20dBm power incidence and at least 18.2% rectenna RF-to-DC conversion efficiency. Self starting is the issue for this topology at very low voltages.

At least 11.3% DC-DC conversion efficiency was recorded by measurement and is comparable to the calculation in Fig. 22. During measurement it was clearly revealed that the boost converter efficiency does depend on the value of L and the duty cycle derived from t_1. To efficiently simulate the complete circuit, from the RF input to the DC output, envelope transient simulation (ENV) in Agilent's ADS was used. The (ENV) tool is much more computationally efficient than transient simulation (Tran). This simulation is appropriate for the boost converter circuit's resistor emulation task. Moreover, the boost converter's DC-DC conversion efficiency, and the overall RF-to-DC conversion efficiency can be calculated at once with a single envelope transient simulation.

In summary, though not capable to operate for voltages as low as 61.7mV DC, the proposed boost converter has by simulation and measurement demonstrated the capability to boost voltages as low as 400mV DC, sufficient for battery or capacitor recharging, assuming that the battery or the capacitor has some initial charge or energy enough to provide start-up to the boost converter circuit.

The limitations of our proposed boost converter circuit include; low efficiency, lack of self starting at ultra low input voltages, and unregulated output. To address these limitations, circuit optimization is required. Moreover, alternative approaches which employ a flyback transformer to replace the boost converter inductance must be investigated. A regulator circuit with Low Drop Out (LDO) is necessary to fix the boost converter output voltage commensurate with standard values like 2.2 V DC for example. For further reading, see [7]

2. Performance analysis of the complete RF energy harvesting sensor system

To demonstrate how one may analyze the performance of an RF energy harvesting system including its application, we extend the discussion of Section 2.5.2 to this Section. We propose a transmitter assembled as in Fig. 26 for temperature sensor wireless data transmission.

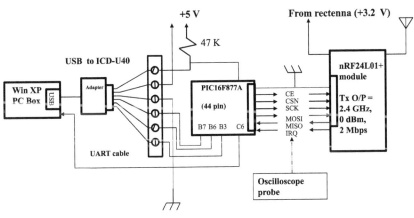

Fig. 26. The assembly and test platform for the proposed battery-free sensor transmitter.

The transmitter consists of one-chip microcomputer (MCU) PIC16F877A and wireless module nRF24L01P for the control, and MCU can be connected with an outside personal computer using ICD-U40 or RS232 cable. The wireless module operates in transmission and reception mode, and controls power supply on-off, transmitting power level, the receiving mode status, and transmission data rate via Serial Peripheral Interface (SPI). Figure 27 shows the operation flow when transmitting.

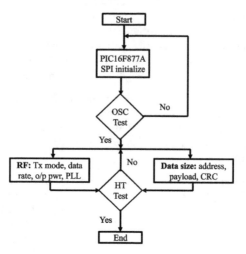

Fig. 27. Operation flow during transmission.

The experimental system composition is shown in Fig. 28 to transmit acquired data by the temperature sensor with WLAN at 2.4 GHz (ISM band). An ISM band sleeve antenna is used for the transmission. Using the cellular band rectenna shown and discussed in Section 2.5.1, at least 3.14 V is stored in the electric double layer capacitor over a period of four hours. To harvest a maximum usable power for the overall system, we charge the capacitor up to 5V. The operation voltage for the wireless module presented in Fig. 26 above is between 1.9V and 3.6V.

The signal was transmitted from the wireless module while a sleeve antenna, same like the one for transmission was used with the spectrum analyzer and the reception experiment was performed. Received signal level equal to -43.4dBm was obtained at a distance 3.5m between transmitter and reception point. The capacitor's stored voltage was used to supply the wireless module in the above-mentioned experiment. Successful transmission was possible for 5.5 minutes after which, the capacitor terminal voltage decreased from 3.16V to 1.47V, and the transmission ended. The sending and receiving distance of data can be estimated to be about 10m when the sensitivity of the receiver is assumed to be -60dBm, given 0dBm maximum transmit power.

Hereafter, the overall system examination is done by environmental power generation using the transmitted electric waves from the cellular phone base station, proposed based on the above-mentioned results. First of all, the power consumption shown in Fig. 29 is based on the fact that 120mW (5V, 24mA) is saved in the electric double layer capacitor by environmental power generation, achieved by calculation as discussed earlier.

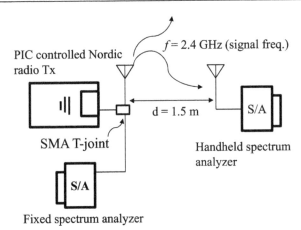

Fig. 28. Indoor measurement setup for received traffic from the sensor radio transmitter.

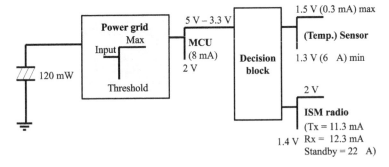

Fig. 29. Power management scheme for the cellular energy-harvesting sensor node.

The sensor data packet is transmitted wirelessly in ShockBurst mode for energy efficient communication. The data packet format includes a pre-amble (1 byte), address (3 bytes), and the payload i.e. temperature data (1 byte). The flag bit is disregarded for easiness, and cyclic redundancy check (CRC) is not used.

The operation of the proposed system is provisionally calculated. When the rectenna is set up in the place where power incidence of 0dBm is obtained in the base station neighbourhood (as depicted in Section 2.5.2), an initially discharged capacitor accumulates up to 3.3V by a rectenna with 53.8% conversion efficiency (presented in Section 2.5.1). At this point, it takes 1.5 minutes to start and to initialize a wireless module, and the voltage of the capacitor decreases to 2V. This trial calculation method depends on the capacitor's back up time discussed in [8]. After this, when the wireless module is assumed to be in sleep mode, the capacitor is charged by a 0.28mA charging current for four hours whereby the capacitor's stored voltage increases up to 5V. The power consumption in the sleep mode or standby is 33μW (1.5V, 22μA).

When the wireless module starts, after data transmission and the confirmation signal is sent, the voltage of the capacitor decreases by 0.6V, and consumes the electric power of 7.4mW.

The voltage of the capacitor decreases to 2V when 3.2mW is consumed to the acquisition of the sensor data, and the operation time of MCU is assumed to be one minute to the data storage in the wireless module etc. As for the capacitor voltage, when the wireless module continuously transmits data for 20 seconds, it decreases from 2V to 1.4V and even the following operation saves the electric power. Therefore, a temperature sensing system capable of transmitting wireless data in every four hours becomes feasible by environmental power generation from the cellular phone base station if we consider intermittent operation by sleep mode.

3. Conclusion

This Chapter has given an overview of the present energy harvesting sources, but the focus has stayed on RF energy sources and future directions for research. Design issues in RF energy harvesting have been discussed, which include low conversion efficiency and sometimes low rectified power. Solutions have been suggested by calculation and validated by measurement where possible, while highlighting the limitations of the proposed solutions. Potential applications for both DTV and cellular RF energy harvesting have been proposed and demonstrated with simple examples. A discussion is also presented on the typical performance analysis for the proposed RF energy harvesting system with sensor application.

4. Acknowledgment

The authors would like to thank Prof. Apostolos Georgiadis of Centre Tecnològic de Telecomunicacions de Catalunya (CTTC, Spain) for the collaboration on the design and development of the DC-DC boost converter circuit. Further thanks go to all those readers who will find this Chapter useful in one way or the other.

5. References

[1] Keisuke, T.; Kawahara, Y. & Asami, T. (2009). *RF Energy Intensity Survey in Tokyo..*, (c)2009 IEICE, B-20-3, Matsuyama-shi, Japan
[2] Mikeka, C.; Arai, H. (2011). *Dual-Band RF Energy-Harvesting Circuit for Range Enhancement in Passive Tags*, (c)2011 EuCAP, Rome, Italy
[3] Pozar, D. (2005). *Microwave Engineering*, Wiley, ISBN 978-0-471-44878-5, Amherst, MA, USA
[4] Mikeka, C.; Arai, H. (2010). Techniques for the Development of a Highly Efficient Rectenna for the Next Generation Batteryless System Applications, *IEICE Tech. Rep.*, pp. 101-106, Kyoto, Japan, March, 2010
[5] http://www.secomtel.com/UpFilesPDF/PDF/Agilent/PDF_DOCS/SKYDIODE/ 03_ SKYDI/HSMS2850.PDF (Last accessed on 13 July, 2011)
[6] McSpadden, J. et al., H. (1992). Theoretical and Experimental Investigation of a Rectenna Element for Microwave Power Transmission, *IEEE Trans., on Microwave Theory and Tech., Vol. 40, No. 12.*, pp. 2359-2366, Dec., 1992
[7] Mikeka, C.; Arai, H. ; Georgiadis A. ; and Collado A. (2011). DTV Band Micropower RF Energy-Harvesting Circuit Architecture and Performance Analysis, *RFID-TA Digest*, Sitges, Spain, Sept., 2011
[8] Mikeka, C.; Arai, H. (2009). Design of a Cellular Energy-Harvesting Radio, *Proc. 2 nd European Wireless Technology Conf.*, pp. 73-76, Rome, Italy, Sept., 2009
[9] Popovic Z., et al., (2008). Resistor Emulation Approach to Low-Power RF Energy Harvesting, *IEEE Trans. Power Electronics, Vol. 23, No. 3*, 2008

Permissions

The contributors of this book come from diverse backgrounds, making this book a truly international effort. This book will bring forth new frontiers with its revolutionizing research information and detailed analysis of the nascent developments around the world.

We would like to thank Dr. Yen Kheng Tan, for lending his expertise to make the book truly unique. He has played a crucial role in the development of this book. Without his invaluable contribution this book wouldn't have been possible. He has made vital efforts to compile up to date information on the varied aspects of this subject to make this book a valuable addition to the collection of many professionals and students.

This book was conceptualized with the vision of imparting up-to-date information and advanced data in this field. To ensure the same, a matchless editorial board was set up. Every individual on the board went through rigorous rounds of assessment to prove their worth. After which they invested a large part of their time researching and compiling the most relevant data for our readers. Conferences and sessions were held from time to time between the editorial board and the contributing authors to present the data in the most comprehensible form. The editorial team has worked tirelessly to provide valuable and valid information to help people across the globe.

Every chapter published in this book has been scrutinized by our experts. Their significance has been extensively debated. The topics covered herein carry significant findings which will fuel the growth of the discipline. They may even be implemented as practical applications or may be referred to as a beginning point for another development. Chapters in this book were first published by InTech; hereby published with permission under the Creative Commons Attribution License or equivalent.

The editorial board has been involved in producing this book since its inception. They have spent rigorous hours researching and exploring the diverse topics which have resulted in the successful publishing of this book. They have passed on their knowledge of decades through this book. To expedite this challenging task, the publisher supported the team at every step. A small team of assistant editors was also appointed to further simplify the editing procedure and attain best results for the readers.

Our editorial team has been hand-picked from every corner of the world. Their multi-ethnicity adds dynamic inputs to the discussions which result in innovative outcomes. These outcomes are then further discussed with the researchers and contributors who give their valuable feedback and opinion regarding the same. The feedback is then collaborated with the researches and they are edited in a comprehensive manner to aid the understanding of the subject.

Apart from the editorial board, the designing team has also invested a significant amount of their time in understanding the subject and creating the most relevant covers. They scrutinized every image to scout for the most suitable representation of the subject and create an appropriate cover for the book.

The publishing team has been involved in this book since its early stages. They were actively engaged in every process, be it collecting the data, connecting with the contributors or procuring relevant information. The team has been an ardent support to the editorial, designing and production team. Their endless efforts to recruit the best for this project, has resulted in the accomplishment of this book. They are a veteran in the field of academics and their pool of knowledge is as vast as their experience in printing. Their expertise and guidance has proved useful at every step. Their uncompromising quality standards have made this book an exceptional effort. Their encouragement from time to time has been an inspiration for everyone.

The publisher and the editorial board hope that this book will prove to be a valuable piece of knowledge for researchers, students, practitioners and scholars across the globe.

List of Contributors

Michael R. Hansen, Mikkel Koefoed Jakobsen and Jan Madsen
Technical University of Denmark, DTU Informatics, Embedded Systems Engineering, Denmark

Chitta Ranjan Saha
Score Project, School of Electrical & Electronic Engineering University of Nottingham, Nottingham, NG7 2RD, UK

Dibin Zhu
University of Southampton, UK

Piotr Dziurdzia
AGH University of Science and Technology in Cracow, Poland

Swee Leong Kok
Faculty of Electronic and Computer Engineering, Universiti Teknikal Malaysia Melaka, Malaysia

Yen Kheng Tan and Wee Song Koh
Energy Research Institute @ NTU (ERI@N), Singapore

Emanuele Lattanzi and Alessandro Bogliolo
DiSBeF - University of Urbino - Piazza della Repubblica, 13, 61029 Urbino, Italy

Luca Gammaitoni, Helios Vocca, Igor Neri, Flavio Travasso and Francesco Orfei
NiPS Laboratory – Dipartimento di Fisica, Università di Perugia, INFN Perugia and Wisepower srl, Italy

Helmut Tributsch
Retired from Free University Berlin, Institute for Physical and Theoretical Chemistry, and Helmholtz-Centre Berlin for Materials and Energy, Germany

Chomora Mikeka and Hiroyuki Arai
Yokohama National University, Japan

Printed in the USA
CPSIA information can be obtained
at www.ICGtesting.com
JSHW011443221024
72173JS00004B/915